QTC

QTC
by "Sparks" (Ray Redwood)

The text of QTC was set in Times Roman on a Hewlett-Packard LaserJet series II. The chapter titles were set in Helvetica Bold Reverse.

Printed in the United States of America.

$15 hard ISBN 0-945845-00-6 SAN 247-8129
$8.95 pap ISBN 0-945845-01-4

Library of Congress catalog number: 88-90627

First printing: August 1989

THANK YOU

Several hundred individuals had enough faith in this writer and in his book to subscribe to it (either by money or "promise to buy") while it was in the process of being written. That process took much longer than expected. Just as Charles II of England was (in his own words) "an unconscionable time a-dying" (1685), the birth of *QTC* was very prolonged (1985-1989)!

In part repayment, this first edition of the book is being presented with a special limitation of 500 autographed copies.

73, Louis!

Ray Redwood

This is copy number #170.

ACKNOWLEDGEMENTS

... to Tom Noe (editor), Terri Nigro (typist), Mark McCulloch (typesetter *et al.*), Stephanie Del Valle (typographer), Todd Green, Norman Annal, Hannah Fischer, Eva Clark, and Paul Ivy (artists), and to the many others who have helped me with this book, I offer my sincerest thanks.

Photos and material marked with the small "M" in the lower right-hand corner are from the Marconi Company, Chelmsford, England, and are reprinted here with permission.

The excerpt from *A Man Called Intrepid* in the chapter on radio in World War II is reprinted with permission of the publisher, Harcourt, Brace, Jovanovich, Inc. The material is copyright 1976 by William Stevenson.

Photos of Marconi, De Forest and Fleming are reprinted compliments of the United States Library of Congress.

International Telecommunications Union General Secretariat, Geneva 20, Switzerland, authorized reproduction of Call-Sign Book on page 345.

SOS to the Rescue (Oxford U. Press, 1935) by Karl Baarslag (an original SOWP member) was the source of the story about *ss Tashmoo.*

"Morse Jargon" is reprinted from "Nautical Magazine" (Glasgow, Scotland), Sept. 1957 by permission of publishers.

My Father Marconi (Degna Marconi, McGraw-Hill, 1962), was the source of many Marconi anecdotes.

Veteran Wireless Operators Association supplied the photos of Battery Park Memorial Heroes from VWOA Yearbooks. Secretary Ed F. Pleuler Jr. was very cooperative and helpful.

Paul du Mesnil (fellow SOWP member, of Nova Scotia) sounded an alarm about the fate of *ss Arctic Explorer* and (with the author) has gloomy forecasts for Century 21.

Authors Gordon Thomas and Max Morgan Witts kindly OK'd much information from *The Strange Fate of the Morro Castle* (Stein and Day, 1972).

Many thanks to Ron Nielsen ("Bookcrafters") for last-minute advice and assistance.

If any copyrighted material in this book has not been credited or acknowledged, the author will gladly make the necessary alterations in future printings, if the correct information is forwarded to him. A diligent effort has been made to locate all copyright holders and gain their permission.

TABLE OF CONTENTS

Thank You iii

Acknowledgements v

Communication (Landmarks/Breakthroughs) 1

Man Goes to Sea 3

Manifesto 5

MARISAT 11

Maxwell/Hertz 14

How Radio Began . . . Marconi Experiments 15

The Originals 22

The Birth of Radio (Part 2) & Henry Jackson 23

Marconi Triumphant 47

Sir Ambrose Fleming 67

Mr. Radio (Dr. Lee de Forest) 69

Some Conventional Diagrams 75

Inventors Responsible for Radio 76

Mr. Electricity (Nikola Tesla) 77

Wireless Aids Man-Hunt 83

Very Early Radio Shacks 87

Again – Very Early Days 88

Early – But Not Quite so Crude – Radio Shacks 89

Broadcast, 1920, Radio Shack 1938 90

Radar 91

Direction-Finding 92

Radio Shack, 1970-1980 93

Some Symbols Explained 95
"Press O Trousers!" 97
A Sparks is Born 99
The Floating Circus 119
Chirrup of Happiness 139
Sumiko 149
Ranges of Eyesight/Earshot 165
Wendens Ambo 167
Hero/Villain George Rogers & *ss Morro Castle,* 1934 175
Stormy World & Heroes Behind a Quiet Little Memorial .. 215
Morse Jargon 227
Long Hair & Trash in Singapore 233
Radio in WW II 239
Grain to Russia 243
Cameos 267
The King of the Radio Shack 271
ss Titanic 273
A Quiet QTP 341
Radio in Lifeboat 343
Call-Sign Book 345
Cupid's Arrows vs. Hammer & Sickle 347
Weather Maps 355
RESCUE! 361
Bibliography 373
Word of Special Thanks 375

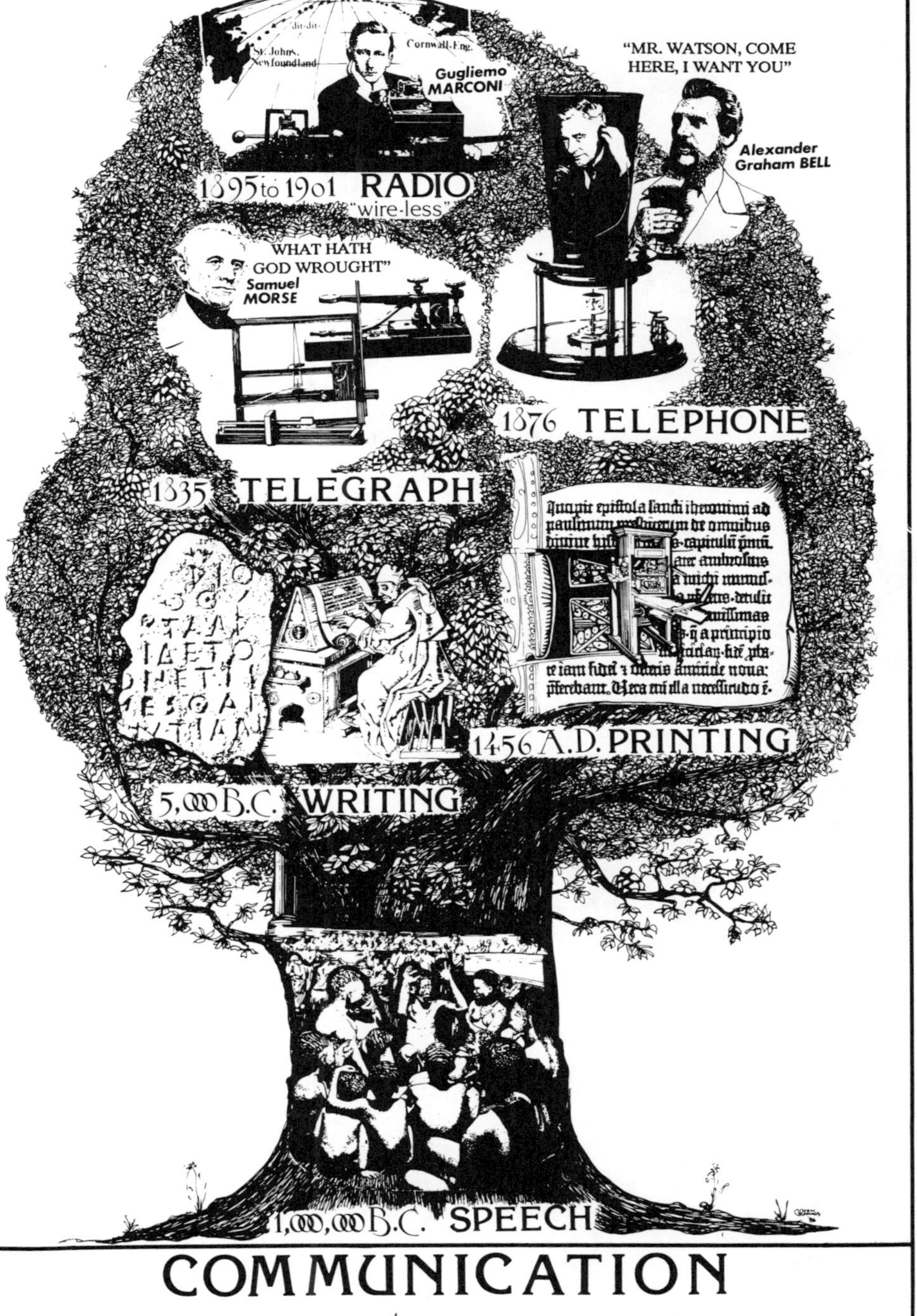

COMMUNICATION

LANDMARKS / BREAKTHROUGHS

Man goes to sea . . .

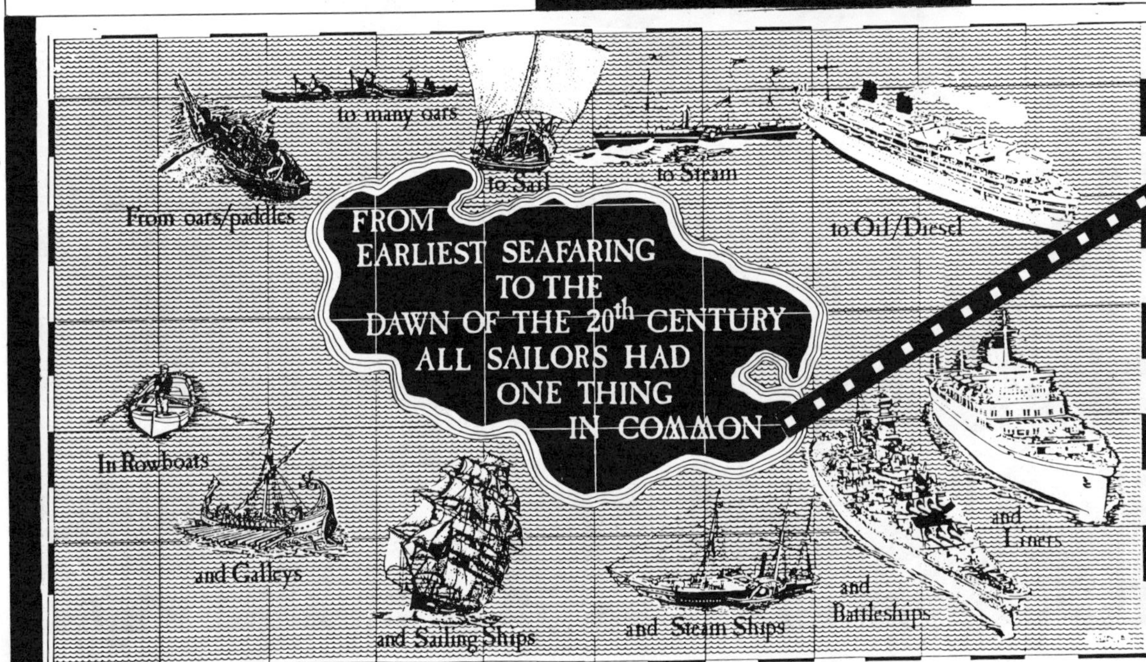

This was . . .

Isolation

Manifesto

A Manifesto is a formal declaration of principles. Something more than a mere preface is needed, for this book. An era is passing.

I am proud and honored to have been a ship's radio operator -- a Sparks -- for most of my working life. No, I haven't been called to the role of Great Hero during some catastrophe. For most of my time I've simply gone about the daily humdrum business of sending and receiving messages to and from coast stations, getting and sending weather reports, exchanging information with other ships.

Even so, the *essential* reason I sailed on each of the scores of ships I have sailed on was simply to be there if I was needed to call for help. Everything else I was paid to do was irrelevant, a mere convenience. If the ship was going down, I was the lifeline for my fellow crew and passengers. That's why I was there. I am proud to have been a Sparks (probably one of the last), and I'll be telling you stories about myself and some of my profession's heroes -- people who were called on in emergencies and went about the task of saving lives, sometimes even at the expense of their own lives.

Today, radio operators are being phased out. I'll explain why, and also why I think it's probably a bad decision. You can decide for yourselves. This book is a tribute to our profession, nearly 100 years old, and to all the radio operators who have joined me on the seas.

* * * * *

Throughout recorded history, sailors went over the horizon into complete isolation. As the 20th century dawned, this isolation ended, through "wireless communication" (later named radio). This communication was based on two important factors: (1) the worldwide use of 500 kilohertz (khz) as the international calling and distress frequency and (2) the presence of Sparks, the man in charge, and therefore the lifeline for his fellow seamen. All other aspects of radio (entertainment, news, etc. -- the only side of it most people know) were later developments, in some ways irrelevant and unimportant.

Thus it went for 70 years. The SOS system on 500 khz, operated by Sparks, was completely successful. Perhaps a million lives were saved by it. In 1970, however, the International Maritime Organization (IMO) decided the time was ripe for a change. Future Global Maritime Distress and Safety System (FGMDSS) was born.

It was internationally agreed that the supremely successful 500 khz and distress frequency system would be maintained for a while, and no innovation made until there had been a lengthy period of testing. Yet some nations jumped the gun (the Scandinavians first, followed by others, including Canada, then the U.S.A.). They freed some of their ships to take to deep sea without radio officers or the usual 500 khz watch -- thereby endangering many lives.

FGMDSS will supersede manual radiotelegraphy formally and completely (maybe) in 1999. After that date, Sparks and all that he stood for and did will be a receding memory. So, on one side we have the lone radio operator, pounding away at his Morse key. On the other we have an automatic communications system that works via control centers in various parts of the world in contact with three maritime satellites varying from 22,314 miles to 22,416 miles above the equator. The new system will eclipse the old one, but will it be safer? Will it always be reliable? Will it do what Sparks did right through the 20th century again and again and again? No, no and no. Ship-owners will save money. "Progress" (which doesn't necessarily mean improvement) will have been made. It's difficult to calculate on suppositions, but I fear that half a million lives could be lost at sea in the 21st century because ships in distress won't be able to communicate with anyone else to get help.

Example: it's not too much to say that the Canadian Coast Guard, among others, has turned a blind eye and ignored the obvious. The fate of the *Arctic Explorer* in July, 1981, is worth noting. She had no 500 khz, no radio officer. While depending on VHF channel 16 and 2182 khz (both voice transmitters/ receivers), she sank in 20 minutes. Out of the 32 people aboard, 13 died and the remaining 19 on a raft (drifting for 30 hours) would also have perished if a helicopter had not accidentally spotted them and summoned a ship for their rescue. Communications via satellites work sometimes, but the simple 500 khz system has been effective 99 times out of 100. The fate of the *Arctic Explorer* points out that high technology is going to fail, where the simple old system would have succeeded. Consider the countless rescues through 70 years. The spectacular example of the *Titanic* -- saving 712 by her radioed SOS -- is well known. The 1,517 who perished would probably have been saved too if the auto-alarm had existed then.

A much lesser known example of how effective the old system was is the remarkable story of what happened on the SS *Tashmoo* on May 7, 1928. She was drifting helpless, completely broken down in the Pacific off Mexico. She had no wireless operator, and her radio apparatus was a burned-up, charred mess. She had started her journey that way, but this was legal. All American *steamers* (the *Tashmoo* was a motor ship) carrying 50 or more passengers/crew had to carry two wireless operators and keep a continuous watch on 500 khz. If the ship carried 49 or fewer -- or if it was a motor vessel -- no apparatus or operator was required.* The wind was blowing a full gale, there was a 35-degree list to port, and the engine room was completely broken down. They were outside the shipping lanes, so no one could see them. With no radio or anyone to operate it, no one could hear them. One crew

---

*Legal loophole for unscrupulous shipowners

member, Arthur Finch, had been a signaller in WW1 and knew the Morse code, but that's all. He'd never seen or used a radio transmitter. He knew nothing about radio (or "wireless," as it was still often called). The transmitter, though burned up and apparently completely broken down, had given shocks to two passengers who had tried to repair it. There also seemed to be a danger of its catching fire again. Still, the captain insisted that Finch should try to make the thing work. Finch tried and failed, tried and failed. Suddenly the motor generator sprang to life! Finch, with no evidence that he was transmitting, with no idea if he was on the right frequency or if he could be heard by anyone, pounded SOS SOS SOS on the key and followed this by the ship's position (handed to him by the captain). His signals were heard by several ships! Eventually, all were rescued.

This is an exciting SOS story, but not incredible. With everything else against them, the survivors were indebted to the simple 500 khz/Morse code system.

So what is this new system, exactly? It's very involved. It's not abstruse; you can understand it as long as you're paying close attention, but it's not foolproof. Here's how a call is made. Instead of hollering SOS SOS on 500 khz, there would be three different ways of calling for help: by high frequency (HF), medium frequency (MF) or very high frequency (VHF). For HF there are five different frequencies to choose from, depending on the position of the ship, atmospheric conditions, etc. On MF there are two different frequencies to use (both at once, if possible). For VHF there are four frequencies. So far we have the option (confusion?) of 11 different ways of sending out the SOS, but we've only just started. (FGMDSS might say, "But there'll be 21 extra, backup chances of getting through." Murphy knows better. His law indicates that if anything can go wrong, it will.)

To respond to that SOS, the IMO has tried to get international cooperation and to organize SAR (search and rescue) teams. This never used to be a problem. A call for help got instant response from all nearby ships, because flags didn't matter a damn. However, once the whole affair became formalized and remote from the ship in distress, going via this and that channel, international disagreements surfaced. Turkey and Greece, for example, have always been at loggerheads. Getting them to agree formally to ways of conducting an SAR: from whom, where and when, etc.--no sir. Then we come to another acronym, DSC (digital selective calling). Instead of just blaring out SOS SOS SOS and maybe not being heard (the radio officer is off watch/auto-alarm is faulty/other mechanical failure), there would be a narrowing down of areas to where help was needed, pinpointing the exact place, distressed ship, and ships to help. If this could be relied on, it would be wonderful, but DSC is as unreliable as SAR. Anyone who has heard the dismal "computers down" refrain from his bank will not be surprised at this.

Computers are magical, powerful things, *as long as there isn't some damn little thing to upset them.* They have one crude need -- electric power. Electric systems can fail. Power on a ship may be reduced to one wet-cell 12-volt battery. Unless/until the magic paraphernalia working into and from the three satellites will operate from just such a humble little battery, all that magic stuff is useless. SAR, DSC and RCC (rescue and coordination center) are all unreliable. Can we find a little acronym that is the key to describing this? LGS (little-girl syndrome) is a candidate:

There was a little girl
who had a little curl
right in the middle of her forehead.
When she was good, she was very VERY good;
but when she was bad, she was HORRID.

Complications are forerunners of failure; simplicity is the harbinger of success. Nowhere is this more obvious than in the contest of Morse code versus voice. Voice gets distorted, misunderstood, garbled with interference, etc. Cassandra was the dismal prophetess of doom in ancient Greece. Looking across the seven seas during the 21st century, she should have a field day.*

Okay, enough about FGMDSS. Re this book: historical facts with personal idiosyncracies, heroes, villains, autobiographical segments, miniature travelogues, love stories--all are part of the stew. On the Titanic story I've decided to go beyond the narrow

---

*See page 12.

confines of radio matters into new theories; who could resist? Anyway, who objects? Anything with a Titanic label on it gets attention. QTC is hard to pigeonhole. If you don't find it interesting enough in one section--turn to another!

I want this book to appeal to EVERYBODY, and for that reason technicalities have been minimized to keep the interest general. Sparks brothers, fellow hams, don't you agree that the great story of Sparks in the 20th century has to be relayed to all?

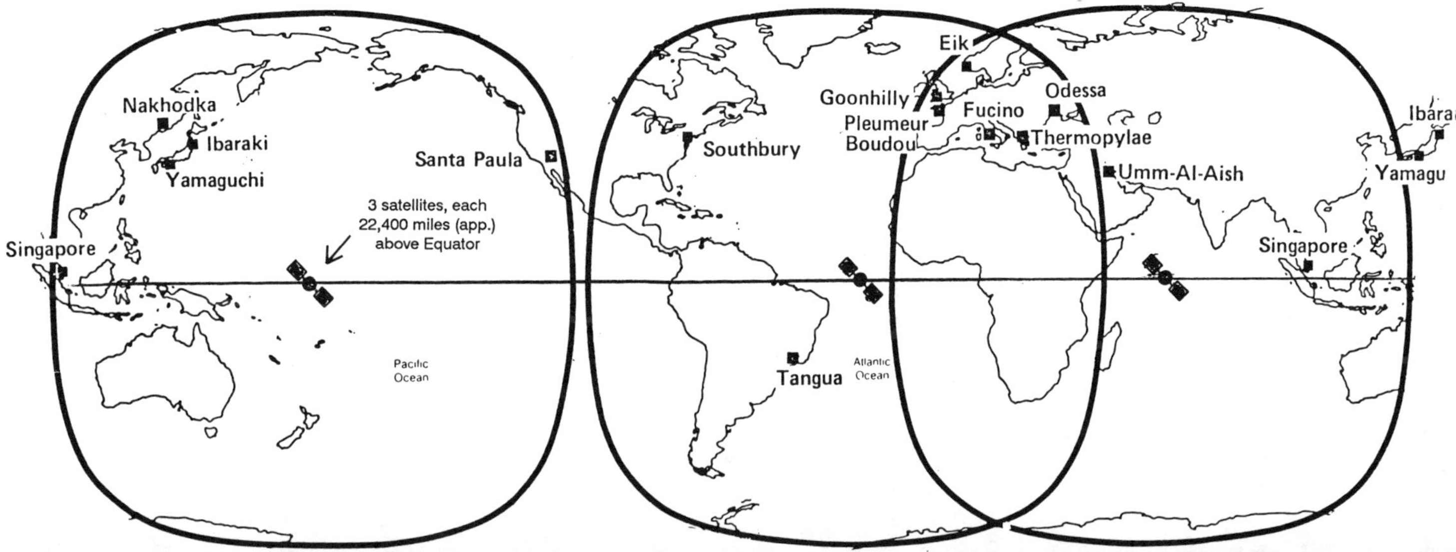

Too far away? Noise? Weak signals? Static? These old bugbears are eliminated. A signal is sent up to the satellite and bounced off it, to return to Mother Earth clear, loud, isolated. MARISAT is the go-between for all kinds of signals: not only Morse but telephone and pictures (facsimiles) of manifests, maps, cargo plans, etc. It's expensive, and it will be a long time (if ever) before every ship can bounce signals off a satellite.

MARISAT (Maritime Satellite)

FGMDSS depends on **satellites**, which may be no damned good in an emergency.

Q. Can FGMDSS equal the simple SOS system?
A. Yes . . . but only IF/WHEN . . .

- All FGMDSS systems working OK?
- Rescuing ship(s) not listing more than 22°?
- Rescuing ship(s) lie between 70'N & 70'S?
- Rescuing ship(s) not in "satellite shadow"?
- Same language used by all concerned?
- Large satellite surface not JAMMED?
- In WW III (or milder war) still workable?
- Delicate ship antennas – weather damaged?
- Gravity of situation communicated quickly?
- The needed telephone circuits not busy?
- Computer has no trouble (major/minor)?
- Computer operator adept/rested/in control?
- All other vessels reporting relevant info?
- . . . & all this data correctly programmed?

All above – <u>ALL</u> – are needed simultaneously!
Omit one of the 14 conditions: system useless!

<u>THIS</u> IS GAUNTLET TO BE RUN FOR SAFETY/SUCCESS!

Nostradamus may already have forecast – but either in addition to him or in place of his occasional lapse – *let this be trumpeted into the 21st century:*

MANY & MANY A SHIP WILL SINK . . .

MANY, MANY, MANY SEAMEN WILL DIE –

BECAUSE { **F G M D S S FAILS;**
500 KHZ IS ABSENT. }

Having signalled the *dangers* of satellites and computers from the vantage-point of 1989, we cannot ignore their *advantages*. These are so great, so far-out, it is not surprising that so many VIPs considered Morse outmoded and crude telegraphy (by wire or radio) could be discarded. The Little Girl Syndrome is true, but . . .

IN FAVOR OF SATELLITES

✓ They're in a trouble-free area, far above the earth and its interference, atmospherics, etc.

✓ They provide nearly instantaneous communication to/from anyplace to/from any ship, via earth stations in USA, UK, USSR, Japan, Norway, Greece, Saudi Arabia and Brazil.

✓ Trials of the original satellite were made in 1976 via the *Queen Elizabeth II* and the hospital ship *Hope.* (The value of high-speed, interference-free communication for medical information and for saving lives is, of course, immeasurable.)

✓ Think, if you will, of the innumerable Sparkses over the past 80 years who have sat in some infamous (for noise atmospherics) area such as the Caribbean, repeating messages again and again and *again,* trying and failing to get through, trying and failing again and again. How like a touch of magic it would be for a fairy queen with a wand to say, "Stop banging your head against a brick wall! Come this way – up, up, 22,400 miles. Touch the satellite and return to earth, trouble-free!"

. . . . . . .But we've been leaping ahead, too far, too fast. Looking at the general situation towards the end of the 20th century and peering into Century 21, we mustn't forget how it all started . . .

James C. Maxwell (Scotsman). In 1873 he foresaw electromagnetic wave communication, wrote a paper on it, read to scientists' meeting in London.

Henrich Hertz (German), in 1886 experimented in his lab with a spark coil — proved Maxwell right. His name lives forever in the terms we use: megaherz (mhz) and kiloherz (khz).

HERZ'S TRANSMITTER 1887

HERZ'S RECEIVER 1887

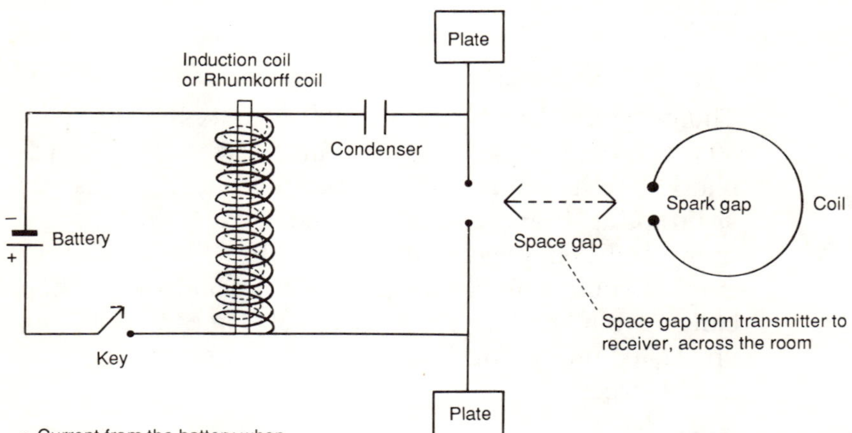

- Current from the battery when key completes circuit.
- It goes thru primary winding of induction coil.
- Current is induced in secondary winding.
- This charges up the condenser.

- When potential difference between spark gap balls is big enough (about 10,000 volts) a spark occurs. ELECTROMAGNETIC WAVES **CROSS TO RECEIVER,** MAKE SPARK THERE TOO.

HOW RADIO BEGAN

Marconi experiments

Radio was born in the fall of 1895 near Bologna, Italy, by the banks of the River Reno. More specifically, it began on the top floor of the Villa Grifoni. This was the home of Guglielmo Marconi, then 20 years old, a young man with no scientific degree, though eventually he was to be given nine honorary degrees. He was, he claimed proudly to the end of his life, "only an amateur."

No scientific achievement is isolated, and Marconi was indebted to many people: Alessandro Volta (1745-1827), who invented the electric battery, Luigi Galvani (1737-1798), who discovered that electricity flows, Andre Ampere (1775-1836), who measured that flow, and Georg Ohm (1787-1854), who gave his name to the unit of resistance and whose law relating voltage, current and resistance was one of the foundation stones of electrical science. In his own century Marconi built on the work of Michael Faraday (1791-1867) of England and Joseph Henry (1797-1878) of America. Even to Thales of Miletus (650-548 BC), the patriarch of all electricians (rubbing amber on your sleeve to get sparks of static electricity began it all!), Marconi owed a "thank you." His immediate sources of inspiration, however, were Professor Righi at the nearby University of Bologna and Heinrich Hertz, who had died at the early age of 37 in February 1894. We'll come back to them; first, Marconi's immediate ancestors have to be introduced.

His father, Giuseppe Marconi, came from landed gentry without much money but a lot of arrogant self-reliance. He had married and soon lost his wife in childbirth, so he was a widower with a small child when he met Annie Jamieson. She was a pretty girl from Ireland with a heavenly singing voice and a hellishly determined will. They fell madly in love. When Annie went home to ask her parents' permission to marry her Italian, the conservative Scotch-Irish Jamiesons were outraged. Absolutely not! He was 17 years older, a Roman Catholic, a widower with a child and -- ugh! -- a foreigner. The pair kept in contact by secret letters, then eloped. They were wed in Bologna, and went to live in his ancestral home, the Villa Grifoni. There, a year later, Alfonso was born. It would be nine years before he had a brother, Guglielmo.

Guglielmo was taught English by his mother. She was a devout Protestant and schooled him in her faith, though he'd been baptized Roman Catholic. Giuseppe didn't care much about religion but was an unyielding martinet in everything else. This built up a reserve in Guglielmo that he retained all his life.

He developed a deep interest in electricity. This was fostered by his mother (who'd have spoiled him if it weren't for stern Giuseppe). His cousin Daisy Prescott was his age, a frequent visitor and a partner in his escapades. His loving brother, Alfonso, also supported him any which way.

Once he rigged a line of dinner plates to electrical wiring and sent them crashing into the garden stream. Another time, after he'd been studying Ben Franklin, he built a contraption to capture electricity from thunderstorms, wired it to a bell and was then delighted to hear the bell ring between thunderclaps.

In other matters he varied. His spoken and written Italian was (he was told by his teachers) vile -- perhaps his Irish mother had confused him. He excelled on the piano, as Alfonso did on the violin. Then, when there was nothing else to do, he could always be found fishing.

Electricity remained his true love. On visits to Leghorn he made friends with a nearly blind old man who had been a telegraphist. From him he learned the Morse code. Here he first tapped out the three dots of the letter "S" -- dit ... dit ... dit; this was the signal he would one day hear by radio across the Atlantic.

The turning point in Marconi's life came in 1894, when he picked up an Italian electrical journal and read about Heinrich Rudolf Hertz's experiments with electromagnetic waves. The article was written by Professor Augusto Righi, whose lectures Guglielmo had attended the previous winter in Bologna. You could say it was also a turning point in the evolution of the world we know and take for granted.

Hertz had experimented with a Leyden jar, a condenser and an induction coil to make a small spark. He strengthened it by a metal reflector. This spark made an oscillatory discharge and started an electromagnetic wave. Marconi, keen to explore along the same lines, read all he could at the University of Bologna, and pestered Professor Righi.

His mother (on his side, always) gave him the run of the villa's top floor, where there were two large rooms. There Guglielmo assembled his apparatus. His simple transmitter was almost the same as Hertz had used. Induction coils provided the power. A primary winding was wound over a soft iron core, and a second winding over the first. Low voltage electricity from a battery was sent through the primary winding, and this induced a high voltage in the secondary coil. This high voltage exploded into a spark between two ball electrodes, radiating electrical energy out in waves.

To record these waves, Marconi used what Edouard Branly (a professor at the Catholic University of Paris) had devised. This was the first crude receiver. When electrical waves passed through a mixture of zinc and silver filings, they clung together, or cohered. So the instrument was called a coherer. It wasn't very efficient. Marconi would improve it.

He got into Professor Righi's classes, probably because Annie Marconi was a neighbor of the professor--and she was very persuasive! However, he couldn't infect Righi with his enthusiasm for the possibilities of electromagnetic waves. Righi was cautious, but at least he let Guglielmo use the university's facilities.

A pause might be worthwhile

A pause in taking for granted the fact (NOW) that no place on earth is more than a tiny fraction of a second away from any other place--and sound/light signals from the limits of the solar

"MR. WIRELESS"
Guglielmo Marconi
1874-1937

system and/or even the edges of the universe are all possible....

Back to that fall of 1894, in September, up in those top rooms at Villa Grifoni and later across the Reno (running sluggishly past apple orchards) where the jagged Appenines shone against the sky. He had locked the doors of the attic (his lab), and repeated experiments he'd already tried in Professor Righi's lab.

The locked door was bait to his cousin Daisy Prescott. When she next visited the Villa Grifoni, she demanded to be let in. At last Guglielmo opened the door, to show her some jars, wires and a wooden reflector--not very interesting.

"You see this needle?" said Guglielmo, showing her a mariner's compass. "Now, watch." He fired up the transmitter on the other side of the room, and the compass needle next to her went crazily round and round. Daisy looked for wires, cords, anything that could have worked this trick (which it must be). She found no visible cause for the needle's behavior.

The excitement spread through the house and reached his father. Giuseppe grudgingly asked his son to show what he'd been up to. Guglielmo's demonstration to his father was to make a distant relay go "CLICK!" when he operated the transmitter.

"That could be a fluke," said Giuseppe. "Make it click three times. "CLICK, CLICK, CLICK" obeyed the relay, and this combination of three dits was henceforth to be Marconi's standard test.

He didn't tell his father that only two days earlier this series of clicks wouldn't have been possible. The chips of zinc and silver had been cohering too swiftly, and would stay that way until Marconi had given the test tube a tap with his finger to loosen them into decoherence. Then he'd had the bright idea of attaching a clapper between the transmitter and test tube so that a surge of electrical power would make the chips cohere, but also operate the clapper to free them again for the next signal.

Still grudgingly, but half-convinced, the elder Marconi agreed to give his son money for more experiments. "You've got to do something a lot more impressive to show me you're not just playing tricks with a toy." The truth of this was bright and hard in poor Guglielmo's mind. He hadn't so far really done anything more advanced than what Hertz had done nine years ago. He'd made improvements, but he hadn't gotten beyond the confines of one room.

The first use he made of the money his father had given him was to buy lots of different materials to experiment for a better coherer. He worked for weeks, and made no less than 500 different combinations until he found that a mixture of 95% nickel with 5% silver worked best. Then he exhausted the air from the tube of filings. Performance in a vacuum was a lot better. In fact, he'd unknowingly just about reached the limits of what a coherer could do. His father's tauntings were driving him to find a really powerful transmitter that would reach out, make contact at a long distance and prove this whole idea of electromagnetic communication worthwhile.

Long waves -- longer and longer waves -- meant progress to him, as it did to Professor Righi, Oliver Lodge, Popov, and everyone else interested in this form of radiation. (It would be some years before the possibilities of short waves were explored.) Experimenting with 30-meter waves, Marconi replaced the two copper balls of the Righi oscillator with two slabs of iron.

Purely by chance, he happened to be holding one slab in the air. The other was on the ground. He discovered that this gave very strong signals. He had accidentally found the clue to effective transmission (it would greatly help reception, too) -- the antenna-ground combination. In some quarters (especially the U.S.S.R.), Alexander Popov is credited with the invention of the antenna. Well, there were simultaneous and coincidental discoveries and inventions during these early days of radio. Independently and honestly each discoverer or inventor came up with his own solution from different angles. It's petty to quarrel about who/when/where/how.

On this occasion, Marconi, by substituting copper wires for one plate (making it into an aerial or antenna) and burying pieces of copper in the ground, had discovered the "Open, Sesame!" to limitless magic communication. In years to come, the symbols for antenna ᛘ and for ground wire ⏚ would be the crude, basic elements in every circuit diagram for transmitters and receivers.

At this point Marconi needed help, and his brother, Alfonso, was a ready ally. Carrying a coherer, Alfonso could get signals from Guglielmo clear at the end of the garden, and waved up to him to signify this. He retreated farther and farther away -- using a long pole with a hanky on the end to mark reception of signals.

Reminiscing much later, Marconi declared, "I knew my invention would have no importance unless it could also communicate across natural obstacles like hills and mountains." (His father's taunts still rankled.) So the next stage was clear. Alfonso had to be sent over the hill, right out of sight. Alfonso, accompanied by two friends (a farmer and a carpenter), took with him a hunting rifle and disappeared over the hill for 20 minutes. Let Marconi say what happened next:

"I started to send, manipulating the Morse key connected to the Ruhnkorff bobbin. In the distance, a shot echoed down the valley!"

* * * * *

That moment, experienced only by the two brothers and their two friends, did not have the spectacular splash of six years later, when three dots crossed the Atlantic to announce to the world that electromagnetic communication was possible by waves following the curvature of the earth.

However, it would be true to say that the answering shot from the valley truly did try to announce to the world and to history that "wireless" was no longer a toy but a worthwhile means of communication for all mankind forever.

It was signalling that the way was open for communication -- not just over the hill, but across the Atlantic, to the moon and back, to the far ends of the universe.

Marconi didn't invent radio. He never said he did. He did invent parts of it, and he was the first to assemble all the elements and make them work together. Many preceded him and some (especially De Forest and Fessenden) followed him soon after, for truly satisfactory results.

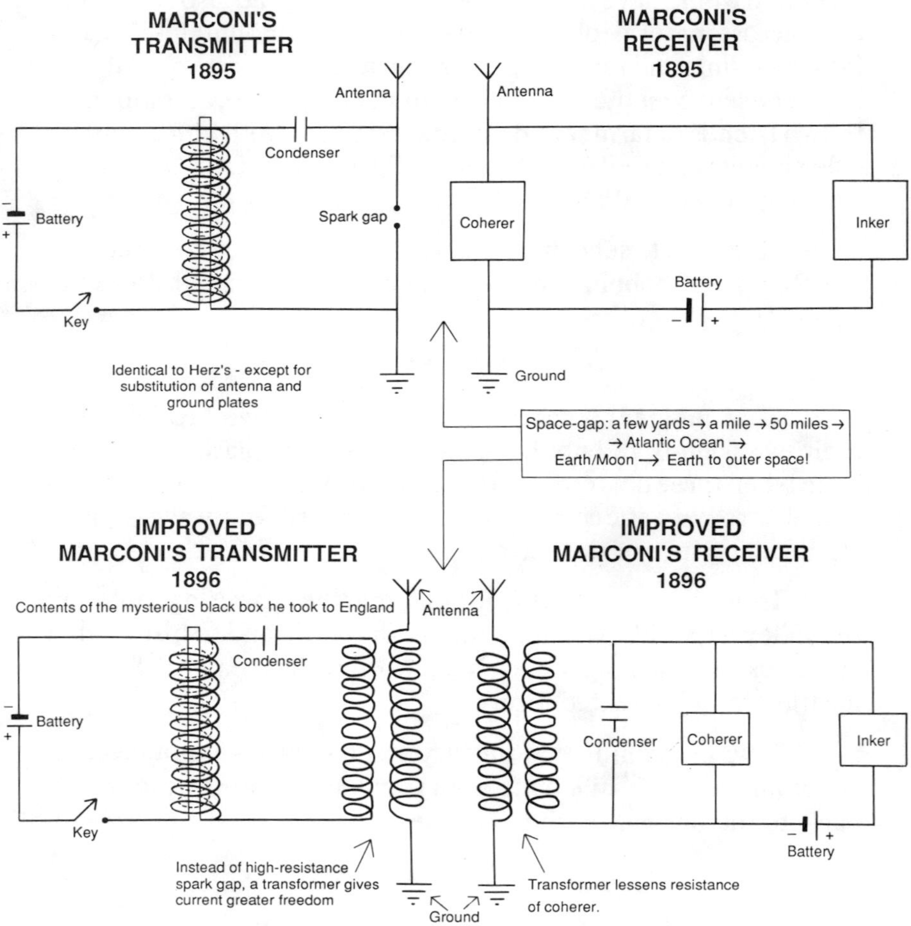

THE ORIGINALS

THE BIRTH OF RADIO

MARCONI (Part 2) with special reference to HENRY JACKSON – A gentleman's gentleman

The exciting tale of radio in its early days, with its heroes and triumphs, also has its villains, shortsighted authorities and sheer skulduggery. Copyright infringements were a considerable problem for Lee De Forest. Edwin Armstrong (another pioneer, who started FM radio) reached the point where he was so worn out by enemies and dead ends that he killed himself.

These shady areas in the bright dawn of radio are mentioned only to contrast with the shining example of Captain (later Admiral) Henry Jackson, R.N. (Royal Navy), a loyal subject of Queen Victoria, then of Edward VII and of George V. His friendship with Marconi and his behavior under pressure from petty-minded folk illustrate his sheer decency, sense of justice and fair play. He was the epitome of the Good Guy.

Marconi arrived in England from Italy in early 1896, rarin' to go. There had always been some mysterious gaps in the historical record concerning his life at this time, because his patron had changed from the Post Office to the Royal Navy. The Navy is traditionally the "Silent Service," and the files on its communications experiments were kept secret. After declassification they were passed to the Public Records Office. They were available to anyone then, but no one knew about them or bothered to consult them until 1972. Then they were unearthed by some gentlemen called Pocock and Grant, who already had some basis for hero worship of Sir Henry Jackson. After 72 years, the gaps were filled. We now know from these hitherto hidden papers just what Marconi and Jackson--separately, then together -- were doing in the last few years of the 19th century. Before we look at these records, though, let's trace their individual stories prior to that fateful meeting of 1896.

Guglielmo Marconi's early experiments developing the ideas of Heinrich Hertz have been described already. Many others in the scientific world were also interested, including Marconi's fellow countryman and neighbor, Professor Righi, and England's Professor Oliver Lodge, who had demonstrated the essential features of wireless telegraphy at Liverpool University in 1894. There was also Professor Slaby, in Germany, constructing an apparatus to make sparks and send electrical power over what had to be called "ether" (because there was no visible connection and sometimes even air was absent). Last but not least, the Russian Professor Popov was as keen as a winter wind howling over the steppes in this worldwide search for communication without wires.

Marconi may have heard of these competitors, but he still went his own way at his family's estate with tremendous drive and inexhaustible patience. These qualities more than compensated for his not being a trained scientist.

He came to England in 1896, partly pushed by his mother, who accompanied him, and partly in reaction against the noncooperative Italian Navy and Post Office. He hoped England would be more receptive than Italy to his real progress in the propagation of electromagnetic waves.

Mr. W.H. Preece, Engineer-in-Chief to the British General Post Office, welcomed him and arranged facilities for exhaustive tests on Salisbury Plain, in the west of England.

Preece -- plus other people and events -- will be dealt with more fully in the next chapter. There's some unavoidable overlapping and a little repetition, as we focus attention first here, then there. Captain Jackson is the main background for now; other people and events (Marconi's womenfolk and more aspects of his career) will follow.

Captain Henry Jackson was there as the Admiralty's observer, and saw signals transmitted two miles, using 13 watts. Jackson was naturally very impressed -- more so than he had been at the War Office a few weeks earlier, when he originally met Marconi. Then Marconi could only show him the room-to-room range of his gear.

Jackson was a dedicated Navy man, through and through. Born in 1855, he was just twice the age of Marconi when they met. He'd been interested in electrical matters since the age of 26 (when Marconi was a *bambino* of five). After starting in the torpedo school, he had been elected as an associate of the Society of Electrical Engineers. His special interest in the propagation of Hertzian waves was indirectly the consequence of his engagement and marriage to the daughter of Samuel Burbery, a mathematician who'd published a treatise on the Maxwell-Hertz theories and practice in 1889. Two years prior to this, some naval maneuvers had brought to light an urgent need. Fast torpedo boats had just been introduced, and the Navy was very anxious to find a way of identifying them, especially at night. Otherwise friendly/enemy vessels could be confused.

This was Jackson's chief interest in the idea of signals without wires. Everything else was incidental or irrelevant. Even so, he couldn't give his undivided attention to the problem because he was always being shunted here and there on naval duties. He couldn't get to the lecture Professor Oliver Lodge gave on coherers, though that was one of his main interests. He had no way of knowing that Marconi was busy on the same things. He may have heard (but probably not) that Slaby in Germany and Popov in Russia were also experimenting with the same material and aims.

All these people, from one end of Europe to the other, were using the same three things: an induction coil (to make a spark and start transmitting a wave); a coherer (a very crude receiver, making small particles sort of lump together); and a bell relay (to register the reception of the signal).

After a lot of shunting, Jackson was suddenly, happily, given command of HMS *Defiance*, an old wooden battleship that looked like Nelson's flagship of 1805. It was a perfect setup for Jackson. He could now experiment to his heart's content, so he did. Beginning in January, 1895, he tried and tried, but got nowhere until December, when his attention was drawn to experiments by Jagadis Bose in Calcutta. It seemed that Bose had been having some success with a toy induction coil giving a one-inch spark, and a coherer comprised of nickel-coated wire springs. These were an improvement on Lodge's tube of metal filings.

Jackson made a replica of Bose's apparatus and started fresh investigations. He made it work, but it needed improvements. In the first three months of 1896, he used glass lenses to focus the radiation of both transmitter and receiver. Again he was successful. Then, in March, he read about Lodge's 1894 experiments and used them as a starting point for new investigations with a two-inch induction coil. He tried different shapes for the spark gap surface. Then he got involved with complicated coherer trials (Marconi did 500 of them, remember?) and finally chose metal filings as the best.

At long last, in July, 1896, he'd assembled a complete receiver. It had a two-foot-long antenna, a coherer of tin and iron filings in an ebonite tube, and an electric bell as a "de-coherer," almost identical to the Marconi arrangement that, unknown to Jackson, had been patented a month earlier. On August 20, 1896, a Morse transmission was successfully sent and received aboard the old wooden ship. In a narrow sense this was the true start of marine radio. (Marconi hadn't as yet done anything on a ship.) Jackson experimented more; he got a range, through wooden bulkheads, of 50 yards. Not much, but a beginning. Two days later he met Marconi.

Both their reputations are greatly enhanced by the story of their subsequent collaboration. Marconi's technical skill and indomitable energy continued as before. As new discoveries were made, he adapted to them, or, rather, he just absorbed them into his tremendous drive. Jackson's behavior was a very pleasant surprise. He commands our respect for his completely unselfish and honorable attitude. Jealousy of his brilliant young collaborator? Desire to capitalize on the other man's achievement? No, these were completely foreign to Jackson's essential decency. Just looking at his face in a photograph of him standing on the deck of the *Defiance*, we see a supremely Good Guy!

Radio pioneer collaborating
with Marconi 1896-1901

Captain (later Admiral)
H. B. Jackson, RN (1855-1929)

He was a loyal friend. Whenever things came to a showdown between Marconi and Jackson's superiors -- usually over penny-pinching economizing, procrastination, ignorance or short-sightedness -- Jackson was there, fighting. He may or may not have guessed that he and Marconi were on the brink of the greatest communications breakthrough in the history of mankind, but he could deal with the fussy little parsimonious objectors who would have brought it all tumbling to dust if they'd had their way.

However, for all Jackson's gentlemanly conduct, there is an ironic aspect to everything at this stage. To appreciate it, we must for a while leave him and Marconi fussing around with wires, coils and interleaved metal sheets.

One of the greatest needs of all seafarers was for land stations and other ships to maintain communication with ships that had gone over the horizon. Countless ships had sunk and all their crews perished because of this lack. (The most tragic were those who had gone just over the horizon -- so near yet so far.) To establish a link with those ships had been at the top of Marconi's aims. It stayed there. Later, his success at having destroyed the isolation of seamen was one of his greatest satisfactions -- right up to his death in 1937.

Here's the irony: These very early efforts to make use of electromagnetic waves were for an armed, death-dealing navy. Jackson was very nice himself, but his interest in radio was only to supplement the powers of the Royal Navy; specifically, as said, to identify torpedo boats.

Marconi's ultimate aim was humanitarian; Jackson's was the efficiency of a fighting force. Their immediate aims and techniques were different, too. Marconi was working on a "beamed" system, using parabolic copper reflectors to focus the waves at both transmitter and receiver. (The goal of this was to get concentration of power.) The Navy needed an apparatus that would (a) be well protected from waves and damp sea air and (b) put up with rough usage--rolling of the ship, and shuddering, heavy shocks, etc. from the firing of big naval guns. Also, instead of the waves being focused, they'd have to be radiated in all directions, so that ships, no matter where they were, could pick up the signals; "beaming" only makes sense if you know exactly where your target is.

After Jackson saw the Salisbury Plain tests of the Marconi gear, he wrote one letter to his commander-in-chief at Devonport, and another to Marconi, explaining all this. Though he was critical of Marconi's apparatus as being unsuitable for ships, its performance (two miles range, against the *Defiance* equipment with a range measured in yards) made it definitely superior to Jackson's. He admitted this, and sought advice from Marconi. During the winter of 1896 and spring of 1897, Jackson redesigned his antennas, making them 70 feet long, and increased his signaling range to 1200 yards. In March, 1897, he got a telegram from Marconi inviting him to observe more trials on Salisbury Plain. Jackson went, and was extremely impressed by Marconi's progress. The maximum range now was seven miles, and signals were readable, in spite of unfavorable weather, at four miles. Always alert to naval requirements, Jackson noted that Marconi's techniques, elevating antennas by kites and balloons, would present little problem on battleships; very tall masts were already there.

He went back to Devonport, got hold of a powerful induction coil to give a six-inch spark, and succeeded in receiving Morse signals from one and three-quarters miles away. This gave him enough confidence to demonstrate his new apparatus to his commander-in-chief. This gentleman was duly impressed and encouraging, but Jackson was appointed to be the naval attache at the British Embassy in Paris, and his further progress in the radio sphere was suspended. Marconi went back to Italy to demonstrate his apparatus to the Italian Navy. After witnessing an exchange of signals between a ship and the Port of Spezia at a range of 16 kilometers (8.65 nautical miles), they were sold on the idea of wireless telegraphy, and set up shore stations at Spezia, Livorno (Leghorn), San Bartholemo and Varugiano. At one of these stations an Italian Navy officer, L. Solari, and a signalman, P. Castelli, devised a new kind of coherer of iron-mercury-carbon that was so successful that Marconi would use it in the first trans-Atlantic transmission in 1901.

Marconi went back to England and formed the Wireless Telegraph and Signal Company. This meant, among other things, that his close relationship with the Post Office was severed; that his doings with the Royal Navy increased; and also that there were financial and business complications very tiresome to tell and to hear. Let's skip them for a while and see the ways other nations, especially navies, were reacting to this new kind of communication.

The French Navy was cool at first, but after the cross-channel communication in 1899, they were enthusiastic. The German Navy sponsored the Professor Slaby system and more or less stayed with that. The U.S. Navy was not in the general "pro" chorus. It had the solitary "con." There was a good reason for this attitude. No transmissions/receptions were, as yet, tuned *selectively*. This needs a little explanation. "Tuning" might refer to getting a set tuned to optimum performance (like a tuned-up car engine). Another meaning is that transmitter and receiver are both tuned to the same frequency. In these early days of radio, lack of it could result in supreme chaos -- any and every nearby signal adding to the confusion. Marconi eventually solved this problem, but in the meantime the U.S. Navy was thumbs-downing radio at sea. What they needed was what was called "syntonic tuning."

The Japanese Navy bought five Marconi sets and used them effectively in their 1900 maneuvers. In 1904, when all their ships were thus equipped, they had a showdown with the Russian Navy and came out on top.

Now that the Russians are back in the picture we can -- must -- examine their claims for Professor Alexander Popov. He certainly had a very early start. His first experiments, in 1887, came almost immediately after Hertz's work. Still, he hadn't made much progress when Lodge lectured and wrote in 1894. He constructed a detector for locating thunderstorms and described it to the Russian Physical Society on May 7, 1895. Independently of him and Jackson, Marconi was working on the same problems with the same raw materials in 1896. All three circuits (Marconi's, Jackson's and Popov's) were modeled after Lodge's induction coil/coherer/bell-relay setup.

In March, 1896, Popov gave another lecture to the Russian Physical Society in Kronstadt, and there is an unconfirmed report that he transmitted a signal 275 yards. In September he was surprised to see a newspaper account of Marconi's activities and realized they were both working with very similar apparatus. He didn't make much progress in 1896, but next year he put his gear on two cruisers, *Russia* and *Africa;* the signaling range was 704 yards. The Russian Naval Ministry, hearing about Marconi's successes, gave Popov more funds. He built better apparatus and got a range of 2.7 miles. This was at the end of summer. (Jackson, in May, had only managed to get one and three-quarters miles, while Marconi, demonstrating in Italy to the Italian Navy, had reached 8.65 miles.)

Popov, experimenting right through the summer of 1898, was working on the same problem as Jackson, but neither knew this. They were both trying to eliminate the relay from the receiver. The eventual Russian solution was to use a telephone receiver, but the *Defiance* apparatus did it with a galvanometer.

In 1900 the battleship *Admiral Apraxin* ran aground in the Gulf of Finland, and Popov set up a radio link, 27 miles long, to help rescue operations. The Russian Naval Ministry was impressed. It decided to install wireless apparatus in all its ships, but took a long time getting around to doing this. In 1901 they achieved communication between two battleships 60.5 nautical miles apart, but only a few ships had radio gear. When war with Japan broke out in 1904, the Navy needed gear badly but the Kronstadt works couldn't cope with the demand. The German Slaby-Arco-Braun Company had to be asked for help. Popov didn't like their sets much.

Thus Jackson, Marconi, Slaby and Popov were all more or less in a line, with Marconi always a little ahead. The U.S.S.R., though (with horse-blinkers on!), sees only Popov's achievements and designated a Radio Day to celebrate Popov's so-called invention of radio.

The final answer to this rather silly dispute comes years later from Popov himself. The king of Italy had made a cruiser available to Marconi for a cruise around Europe, through the North Sea, into the Baltic and on to Russia. There the king of Italy happened to be visiting the Czar and both monarchs came aboard to see Marconi's apparatus. They were followed by Alexander Popov. He came up to Marconi and declared with enthusiasm, "I welcome you and salute you as the Father of Wireless Telegraphy!" This little story (suppressed and denied by the U.S.S.R., of course, but true and well documented) is as refreshing as cool, pure air over a swamp and gives us added respect for Popov as an honest man.

In this mood we're also reminded of our Good Guy Henry Jackson. Back to him. As mentioned, his other duties kept interfering with his radio experiments. When his superiors decided they needed an experienced torpedo officer at a top-level meeting with the French, he had to pack his bags and leave the *Defiance*. He left behind a trained staff, including a Captain Hamilton, who was very familiar with past experiments and future intentions. Above all, they would be paying attention to all his friend Marconi's activities. Jackson went off to Paris with a light heart.

Marconi certainly was in the limelight now. After his formation of the Wireless and Telegraph Company and weakening of links with the Post Office, he astounded some naval bigwigs by demonstrating signals sent between Bournemouth and the Isle of Wight, a distance of over 14 miles, using antennas 120 feet high. In 1898 this was the farthest contact to date by anyone, anywhere.

How had it been done? Marconi and Jackson, working together, had found the clue -- a coupling transformer, a vital link between the transmitter and its antenna. They had delightedly nicknamed it the "jigger." The jigger was to be a magic open-sesame to long-distance communication.

That was only the first of Marconi's spectacular feats in the next few years. His next was the result of installing radio on a ship (first time ever) -- the royal yacht *Osborne*. This was to keep Queen Victoria informed about the progress of her son Edward (future King Edward VII), who was convalescing after a knee injury. About 150 messages passed during August, 1898, linking the two coasts of the Isle of Wight. This wasn't a big distance, but it was spectacular because it had royal patronage and was in the newspapers--very welcome to the newly born Marconi Company. Also, it showed the reliability of routine wireless messages on a seagoing vessel. Twenty-four-year-old Marconi was a public figure in Britain.

Barnum and Bailey couldn't have devised a more spectacular feat than his next, linking England and France by cross-channel communication. He could see few if any technical difficulties. He'd just gotten perfect communication over 14 miles; the obstacles might be political only. So it was nice having Jackson in Paris. He wrote to him. Jackson, strangely, was not encouraging. He, too, expected red-tape difficulties unless (1) the French and British Post Offices got together and (2) Marconi moved, with all his instruments, to France for further experiments. "But," he went on, in his replying letter to Marconi, "why not try from the Isle of Wight to Portland Bill, about 45 miles? It would show the possibility of the channel scheme."

Their hesitancies were groundless. The French government was very willing to have Marconi establish a station in France to get a cross-channel link, but they preferred it to be in the Calais area. Marconi had wanted it near Cherbourg, closer to his Bournemouth transmitter. Calais was too far to reach and he hadn't yet got a Dover station established, so the cross-channel scheme had to be shelved during the summer of 1898. Marconi didn't fancy the Portland Bill-Isle of Wight project suggested by Jackson. He was in Show Biz now! The cross-channel transmission would shout to the world. He was going to erect a station in Dover, to contact Calais.

This wasn't ready until November. While he was busy in this southeast corner of England, he was approached by the Trinity House authorities (the Coast Guard of the UK). They suggested he set up a wireless link between the South Foreland lighthouse and one of the three lightships guarding the infamous Goodwin Sands (where many and many a ship had foundered). Marconi agreed, and chose to equip the East Goodwin light vessel, which was 10.4 nautical miles from South Foreland. The crew and her master were taught how to operate the apparatus by a Marconi assistant. They caught on quickly (two days) and he left them in complete charge. It was now January, 1899.

On the second of March, the French government gave Marconi permission to establish a station on the French coast. He and his assistants were waiting and ready. They took apparatus over and established it at the Chalet D'Artoise Wimereux near Boulogne. They aimed at the South Foreland lighthouse, and successfully sent a signal to it at 5:00 PM on the 17th of March--a distance of 32 miles.

Direct link to Europe without wires!

This inflamed the public's imagination as nothing else had so far done. It caused pessimistic forecasts in sober minds, too. Cable shares fell heavily on the Stock Exchange; land lines and submarine cables seemed doomed. Then, after an initial panic, it was realized that wireless telegraphy would complement, not replace, cables.

Christmas Eve, 1898. First wireless on a ship, ever.*.first calling device to get help. Lighthouse at South Foreland (guarding Goodwin Sands, at mouth of English Channel). Man in center has hand on Morse key. One touch of it rings bells ashore, 12 miles away. This was done when ship ran into lightship, March 1899: message was sent, lifeboats came. Nothing similar nor so effective until auto-alarm came into use nearly 30 years later.

*Well, not quite! Royal yacht had wireless in August 1898.

On the 28th of April, 1899, the wireless set on the East Goodwin light vessel, after only three weeks in operation, proved its value to the hilt. The steamship *R. F. Matthews*, looming out of an early-morning fog, struck the lightship--not quite head-on, but enough to cause severe damage. Captain Clayton called the Marconi operator at South Foreland by picking up the phone (connected to radio gear, not a landline) and sent what was the first radio emergency signal.

"We have just been run into by the steamer *R. F. Matthews* of London. Steamship is standing by us. Our bows very badly damaged."

Lifeboats were alerted, but not needed. No one had been hurt, in the steamer or light vessel. Still,

(1) a historical "first" had occurred;* and

(2) for the second time in a month Marconi was headlined.

You may remember that when Marconi had just come back to England from Italy (while Jackson was being posted to Paris), you were given hints of trouble, petty wrangling, etc., ahead. Well, recent triumphs (cross-channel and Goodwin vessel) had not eliminated this trouble spot. It involved the Royal Navy and, much against his will, Captain Jackson. The Navy, into which Jackson had funneled his enthusiasm for the Marconi apparatus, was anxious to give it extensive trials, especially during the naval maneuvers of 1899. Remaining technical problems were few. Most had been smoothed away by the Marconi/Jackson team. Still, there were administrative problems and outside factors, centered in Marconi's connection with the Post Office.

He'd been sponsored by the Post Office since coming to England, but he'd not received any financial help. It was all informal and friendly. Now that he had his own company -- on the

---

**Sounding an alarm -- to provide instantaneous awareness of the need for help. Here is was being done perfectly, when radio had only just been born! It was needed 13 years later, when the* Titanic *was sinking, but wasn't available. The auto-alarm system didn't become really efficient and available until 29 years later.*

Stock Exchange, too -- there had to be changes. A break had to be made.

The uncertainty about money/contracts/grievances, etc., made for an unfortunate atmosphere, and the Navy didn't want to butt in while another government department was wrangling over general policy and minor details. On the other hand, they were getting impatient.

Someone had a bright idea; why not speed things up by questioning the validity of Marconi's patent? Jackson had been experimenting as long as Marconi, and had built some sets almost as good as Marconi's. Maybe they could cut Marconi out of the picture entirely, and steam ahead with Jackson's stuff. Captain Hamilton (who was Jackson's temporary relief on the *Defiance*) and Jackson himself were told to try to invalidate Marconi's patent by evidence of what Jackson had previously done.

This sneaky plot might have succeeded if Jackson had cooperated (he didn't, of course) and if Captain Hamilton hadn't also been an honorable man and loyal friend of Jackson. They wrote separate letters to the VIPs in the Admiralty saying (via formal gobbledygook) that Marconi's patents were solid. He'd anticipated *all* Jackson's discoveries, and anyway they better stop nitpicking about who started what and when, and give this vital, wonderful means of communication a thorough test in the upcoming maneuvers.

So the Admiralty, grumbling and half-willing, started negotiations with the Wireless Telegraph and Signal Company to get their apparatus installed on two ships of the fleet. Then, when the Board Meeting got their first glimpse of the royalties the company wanted, there were loud objections that they were too high. Marconi was asking the Navy for £100 per ship per annum. By modern standards this seems absurdly low, but the standards of the 1890s, plus the estimated total cost for the entire fleet, made that cost seem excessive. Then their lawyers entered the picture with plans for an attack on Marconi.

The government had a monopoly on telegraph signals within the U.K., so why not quote this as a sound reason for getting this valuable invention for nothing at all? National defense was at stake; Marconi couldn't be allowed to make money on such a "delicate" matter.

Meanwhile, Jackson was ending his tour of duty at the British Embassy in Paris and was being sent back to wireless experimentation in a big way. This memo came from the Director of Naval Ordnance:

"... It is understood that Captain H.B. Jackson is to be in command of a ship during the forthcoming maneuvers. Wireless apparatus will be used on that ship...."

His joy must have overflowed when he learned, in addition to the above, that Marconi was going to be on board with him, supervising, observing, helping.

For lack of evidence, we must use our imaginations to reconstruct their rendezvous. Marconi has arrived in Milford Haven, the big naval depot on the southwest tip of Wales. He has taken a launch out to the cruiser HMS *Juno*, where his friend Jackson is waiting for him. At the top of the gangway they beam at each other, shake hands very warmly, and proceed up to the captain's cabin. It's the 12th of July, 1899; they haven't seen each other for eight months. They settle down in the lavish office assigned to the ship's commander. Jackson, British navyman to his core, has his rum and soda water handy, but so far he's been unable to tempt Marconi to forsake his campari and soda.

"Well, Bill," says Jackson (Guglielmo is Italian for William), "the door's closed. If anyone heard what I think of those folk in Whitehall, I'd get a severe reprimand! Just tell me, how are things going with my bosses and your company?"

"Not good, Harry. How do you say in English? They cannot see beyond the ends of their noses ... but we're going to *astound* them! Look, we're going right out into the Atlantic, yes?"

Jackson, smiling, just nods. He doesn't want to interrupt this boiling enthusiasm even for a second.

"Those jiggers! -- best thing we ever did -- they're going to stretch the range maybe three, four, five times. I've been testing them, but not at sea. When there are no hills, buildings, nothing to interfere -- well, we'll see...."

Marconi had underestimated. The range was 10 times better. The "B" fleet Jackson commanded had intercommunication extending not just the seven miles one could expect on land, not just the 28 miles they'd once got with jiggers, but 70 miles. Previously, to get a message across that distance, using flags, light signals, ship-to-ship relay (the overriding need being that they had to be within *visual* range) had taken as long as 55 hours (instead of a fraction of a second, with the wireless).

The triumph of wireless telegraphy was greater than words could describe. As a *fait accompli*, it was a silent slap of the obvious. Marconi and Jackson vigorously pumped their right hands together and pounded each other's chests in joyful exuberance.

How would those wet miseries in the Admiralty react? Jackson, in his official report, proposed that wireless apparatus be established throughout the fleet, with a trained signalman at each point. He also recommended the formation of schools, with a syllabus drawn by Marconi. His immediate commanding officer backed him up, saying, "The system of wireless telegraphy is absolutely invaluable." Consideration of the matter went up through the successive echelons of the Navy. The Third Sea Lord approved. The Second Sea Lord added that at least eight sets should be purchased immediately from the wireless company. The First Sea Lord (the fellow in charge) agreed with the others: "But ... (uh-uh -- here it comes) "... the royalty asked for was excessive, and should not be considered as binding the Admiralty for any future agreement."

The first monkey wrench had been thrown in; the second one came swooping down from the Director of Naval Ordnance: "The Post Office has not yet finished their attempted bargaining with Marconi and his company. The way is not clear for the Navy to bargain."

So Marconi was almost back to square one.

He should have been coasting downhill in glory, considering his thumping successes so far. Despite all this, though, both the Navy and Post Office were unfriendly. His company was in severe financial trouble (over two years old, it hadn't a single steady customer, and there were no dividends for its investors). The scientific world was divided on whether he had the right to make money out of a monopoly on the Hertzian waves. He had rivals and enemies galore.

With so much stacked against him, he thought it was a good idea to get out of the country for a while -- go to the U.S.A. to report on the America's Cup races. He'd been commissioned to install wireless gear on one of the steamers following the yachts. It performed faultlessly, sending thousands of words for press reports. He had hoped to combine this almost social visit with a business deal aimed at the U.S. Navy. Perhaps their initial interest could be fanned into a firm commitment and a big sale. His hopes for this were very high when he had successfully kept two battleships in good contact 36 miles apart. Yet -- no go. The U.S. Navy was going to be even more ornery than Her Majesty's one! They wanted signals to remain intelligible when (1) the tall buildings of lower Manhattan (not really skyscrapers in those days) were in the way; and (2) when shore stations were transmitting at the same time -- in one word, jamming. This was the untuned circuitry referred to earlier. The Marconi sets couldn't pass these tests. Marconi already had devised a way to overcome them, but since he hadn't yet gotten the patents, he held off revealing them to the U.S. Navy. They held off their approval, too! Marconi went back to England with no contracts.

There he found the situation had gone from bad to worse. The Director of Naval Contracts summed up his wireless company's rental terms as "preposterous!" The Admiralty lawyers had figured out that, since wireless was needed for national defense, they could override Marconi's patents and go right ahead to make their own transmitters and receivers. For this they'd need Captain Jackson....

Oh, was he unhappy! He was being forced to be disloyal to his friend by orders he couldn't disobey. He had already designed some transmitters and receivers that were nearly as good as Marconi's. It seemed as if the Navy would use them exclusively. They were making the relationship between himself, Marconi and the wireless company very awkward indeed. He wrote to his superior:

> "As I have had much personal intercourse with Mr. Marconi, I should very much like that he or the company be informed of the steps the Admiralty intend to take, i.e., if the manufacture will be independent of the company, as in that case I shall feel bound to hold no further correspondence or intercourse with him. I would observe that I have refused one invitation of the company to visit their Chelmsford Station and Works since this matter has been in abeyance."

The Admiralty told him, told the wireless company (via them, of course, they told Marconi himself) and told the Post Office, "We're making our own sets." They did just that. They experimented on several ships. Sometimes the results were good, sometimes not. Faulty relays, antennas and other technical faults kept cropping up. Trials went on, but they were forced to conclude that the Jackson sets were not as completely reliable as the Marconi ones, so the high-and-mighty Royal Navy had to climb down. How to save face? Informal discussion was arranged between a rear admiral and the managing director of the wireless company and -- oh, it's too tiresome to go into sticky details. All that matters is (a) the wireless company got the exact contract they'd been angling for, and (b) the Marconi-Jackson friendship was completely open again and as thick as ever.

There's one very interesting government paper that came to light not too long ago, dealing with this period of 1899 to 1901. (Those papers that had been hidden for 72 years, remember?) It is the complete list of all the ships in the Royal Navy scheduled to receive wireless telegraphy apparatus, 31st December, 1901. All the radio sets are identified as to their origin -- whether from Marconi or from Jackson. There were 57 ships with Jackson sets assigned to them, versus 29 to get Marconi sets -- almost exactly two to one! It's an official government paper, listing how the wireless apparatus was distributed. Looks as though the Navy didn't really "climb down" as much as they promised. This list is a complete explanation as to why so many Navy men were honestly convinced that Jackson was the inventor of radio.

Marconi, always in the public eye, went from triumph to triumph. The British-humor magazine *Punch* kept track of his progress (as the 20th century got into stride) by cartoons and drawings very much in his favor. For instance, there was one on October 22, 1913, just after the Volturno disaster (a tanker on fire in the Atlantic -- 521 passengers and crew were saved by the radio's SOS). The cartoon shows Mr. *Punch* saying to Marconi, "Many hearts bless you today, sir. The world's debt to you grows fast." The landmark for all time, of course, is the Titanic disaster of the year before, in which 1,517 died, while 712 were saved by radio.

Jackson also went up and up the ladder of triumph, but wasn't publicized like Marconi. He was promoted and promoted until, just as the Great War was starting, he had nearly reached the topmost rung -- as the First Sea Lord. When he got there, in 1915, the only one above him was the king -- who had no say at all in the running of the Navy. If Lord Henry Jackson, R.N., wanted anything, he got it. What did he want now? Anything and everything connected with (a) his beloved Royal Navy; and (b) the best tool he could find for the Navy -- what we now call radio-communication. He recalled those tortured days at the turn of the century when (with Marconi) he'd had to argue and plead with the top dogs of those days -- how sweet it was to be the top dog now!

He and Marconi, despite their buddy-buddy personal relationship, and despite their common interest in radio, were poles apart in their ultimate aims. Marconi's was to end the isolation of seamen and to save lives. Jackson's was to have the utmost efficiency in a fighting service by having close communication; regrettably, this might mean taking lives, to achieve victory.

Lord Henry Jackson, R.N., commanded the world's most powerful navy. The second strongest belonged to Germany, with whom his country had been in mortal combat since August, 1914. The German fleet wouldn't face his; it was skulking in the harbor of Konigshaven. There were many ships there; what Kaiser Wilhelm or his admirals wanted to do with them was uncertain. They might come out for a showdown in the North Sea, a sort of world's heavyweight championship with the Royal Navy, or they might try to sneak out in the night and go down the English Channel to reach the broad expanse of the Atlantic. They'd be hard to find there, and they could join their U-boats in hunting down Allied merchantmen. In that case, they'd have to be sudden, secretive, and scurry down the English Channel before the British knew what was happening or had time to muster their Navy into readiness. Whichever plan was chosen, timing was of the essence for both Jerry and the Limeys. Jackson had to know if -- *and when, to the very instant* -- the German fleet started to leave Konigshaven (usually called Wilhelmshaven).

He knew just how to get this vital information. His experiments for years with the Hertzian waves had included direction-finding. The direction from which signals were coming to a certain spot could be plotted as a line on a chart. Then the direction of those same signals to another spot could be plotted--and where the two lines crossed was a "fix," the origin of the signals. Direction-finding (D/F) is a wonderful way for ships to find out just where they are from many miles away, even in darkness or fog. Its use to Jackson, however, would be to indicate movements--any movements--of those land-locked German ships. For pinpointing such movements exactly, he had three D/F stations on the east coast of England, all trained on Wilhelmshaven. (See diagram, page 43.) He had those stations manned 24 hours a day. Any suspicious evidence had to be relayed to him immediately, wherever he was. So that meant either his office in Whitehall or his home on Hayling Island on the Thames. He waited.

He was waiting on May 17, 1916. He'd been waiting since August 6, 1914. From the few who knew what the setup was (his friends had hushed it up, out of embarrassment) he could expect no sympathy. The English have always had a good-humored acceptance of their eccentrics, especially if no one is being harmed by them, but to expect any positive results from this crazy scheme involving machines on the east coast pointed toward a spot on the German coast, 200 miles away....

Then it happened, at 2:38 AM on May 18, 1916. There was a tiny movement detected. It was so tiny -- a mere one and a half degrees -- that the naval personnel involved were very reluctant to call Jackson. "They could just be moving one ship for repairs."

"I agree -- let's wait a while, see if it goes on."

"He certainly won't like being awakened at 3 AM to be given a red herring."

"Shut up, all of you! *Call* him! He has always insisted on this." He was called.

Jackson stood in his pajamas, holding the separate phone pieces in his hands, without the slightest doubt in his mind. The long time he'd been waiting, the accumulated cynical doubts of others--all these were just a measure of his *faith* in the evidence.

He telephoned all the right people, for action. By dawn, just about every ship in the huge British fleet -- battleships, cruisers, destroyers, even the torpedo boats (the craft for which he'd started investigating these Hertzian waves) -- *all* were heading east at top speed.

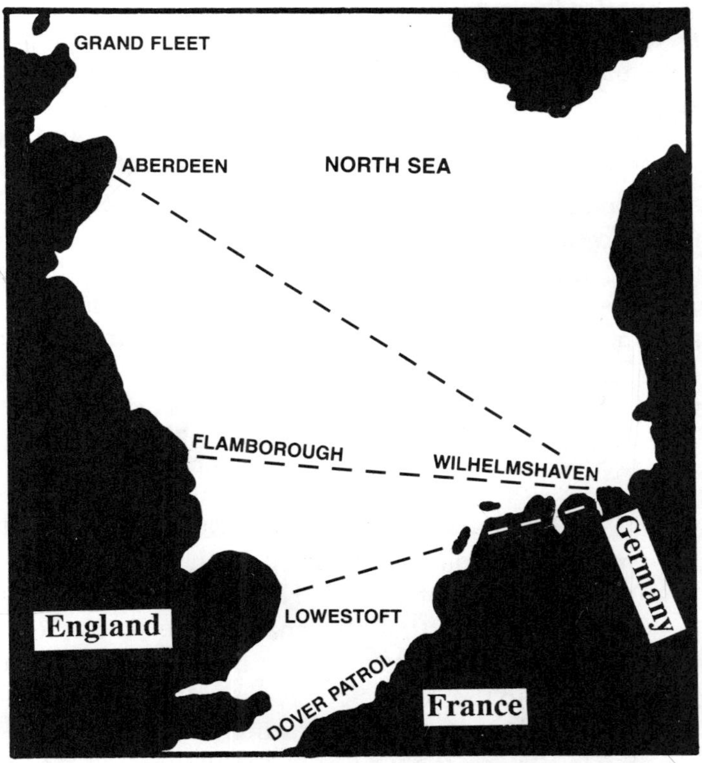

Position of DF (Direction Finding) stations monitoring movements of German fleet.

There were direction-finders on these ships too -- and they could detect the German fleet from 60 miles away ... 55 ... 48 ... 25 ... 20 ... 15 .... Then they could be seen by naked eyes. All guns, British and German, had been loaded, pointed, ready ... "Fire!" "Fire!" "Fire!" "Fire!" "Fire!" "Fire!" "Fire!" Both sides were echoing themselves and each other, deck to deck. Ship to ship. The Battle of Jutland had begun.

The carnage was dreadful. Similar wholesale slaughter had been taking place, 150 miles farther south, amid the fields of Flanders and France. There, it would go on for another 30 months -- all to dispute a few hundred yards of muddy trenches, the so-called front line. The situation in the North Sea, when the smoke of battle had cleared away, seemed also indeterminate. Who had won? Judging by the ships and men lost, it seemed as if the British had been harder hit, but this turned out to be only a tactical defeat. Strategically, the Germans were worse off. Their fleet crept back to Wilhelmshaven like a wounded, whimpering lion -- and did not emerge again for the rest of the war. Britain had a naval blockade of Germany without any fear of interference from the German fleet. Then, when the war ended on November 11, 1918, the German Navy had its supreme humiliation. It was ordered to leave Wilhelmshaven (this time escorted by the British fleet), sail up to Scapa Flow, north of Scotland, and there all its ships had to scuttle themselves. Two years and seven months after Jackson (shivering in his pajamas at 2:45 AM) had been told that the German fleet was emerging, it was rusting at the bottom of the North Sea. The beginning of the end had been the evidence (on a D/F screen) of one German ship (in Wilhelmshaven) moving one and a half degrees!

In 1919 Jackson retired. His naval career had just come to a fitting, typically triumphant conclusion, but he treated retirement as if it was a beginning, not an end. His energy was amazing. He was appointed to be the first chairman of the Radio Research Board, a government undertaking. He never missed a meeting. He'd been elected to the Royal Society since 1901, a great compliment to him as a scientist at that time, when he was only an up-and-coming captain. There, too, he would stay very active. In the last decade of his life he was a big noise in many professional organizations, nearly all of them -- need it be added? -- connected with radio.

He died at 74, as he'd lived: a happy man, with his radio-pioneer boots still on.

To all who knew him, he was a perfect gentleman and an officer of the highest caliber. In all his dealings with and involving Marconi, he never exploited the situation for his own advantage. He could have, easily. The Navy was on the point of using his sets exclusively, but he wanted the best for the Navy, and Marconi's sets were just a little bit better. We could do with a few more Henry Jacksons!

MARCONI

TRIUMPHANT

*Firm achievements

*(but shaky love life!)

*Worldwide fame

We've sidetracked a little in the Marconi story to cover Henry Jackson. Let's go back to when they met -- and earlier. There are dividing lines, turn-off points, Y-junctions in everyone's life. In Marconi's,

ALFONSO'S RIFLE SHOT

was the most important landmark of his early life.

It changed Giuseppe Marconi's attitude completely. He tried (on behalf of his son) to interest the Italian Postal Telegraph, but failed completely. He should have tried to interest the Navy. Anyway, the Italian government rejected his offer, refusing to underwrite Guglielmo's work.

Italy's loss was to be England's gain. Marconi's mother Annie wrote to her nephew Henry Jamieson Davies. He said he'd do what he could if/when they came to London. In February, 1896, Guglielmo and his mother set out for England. His precious apparatus was in a black box, which he had packed, locked and carried himself. The suspicious English Customs opened it, found wires, batteries, dials, weird-shaped pieces of metal. These odd things caused excessive alarm. Queen Victoria had been shot at, three separate times. The French premier had been murdered by an anarchist only two days earlier. There was obviously only one thing to do -- they smashed all Marconi's apparatus beyond repair!

He and his mother arrived in London in low spirits, but cousin Henry (who met them at Victoria Station) bundled them into a horse-drawn cab. Marconi was soon assembling stuff to replace the smashed apparatus. Annie found them a house in the respectable neighborhood of Bayswater, and there Guglielmo feverishly started preparing his patent papers. Cousin Henry got him a good patent lawyer, and he officially deposited his patent papers in June, 1896. He then sought opinions from people of public authority.

The British Post Office, then as now, was in charge of everything to do with communications. England had been queen of the seas since Nelson's time, and one continuing problem was alerting lightships of approaching storms to prevent shipwrecks. Previously mentioned William Henry Preece had been working on the idea of wireless himself. Guglielmo went to see him. The gentle

63-year-old Welshman and the young, nervous Italian got on well together.

Marconi set up his apparatus, made it work, and impressed Preece, who was his staunch colleague thereafter. He let Marconi use his own lab, and even let him take one of his most valuable assistants, George Stephen Kemp. Red-head Kemp became Marconi's best assistant for over 30 years. Marconi's demonstrations through and across rooms in the GPO were witnessed by Captain Henry Jackson and experiments continued out on Salisbury Plain, near Stonehenge. The range of transmission/reception was extended eventually to nine miles. When, in 1897, Marconi sent wireless signals across the Bristol Channel, the world was watching.

That same year, Marconi returned to Italy at the invitation of her Navy. He made a great impression on King Umberto and Queen Margarita. His tests (moderate success) on an Italian warship were a great triumph for the boy who couldn't even qualify for the University of Bologna.

After he returned to England, and had established his own company in 1898, he concentrated on trying to perfect ship-to-ship and ship-to-shore communications. His success, with Jackson, has already been described. This innovation would result in countless lives saved and cargoes preserved. He'd be prouder of this than of his many other accomplishments.

Poole (near Bournemouth) was to be his headquarters for 28 years. His wireless was tested on ships, up and down the coast, usually in very bad weather.

In 1898, he reported on boat races from Dublin. Then Queen Victoria was worried about her 57-year-old son Prince Edward, who had wrenched his knee. He was convalescing on a ship going around the Isle of Wight. Marconi kept mother and son in contact--and got himself some excellent publicity.

About this time, Marconi and Queen Victoria had an amusing confrontation. She was 79. He was 24. They were both pigheaded. Well, he used to take a shortcut to his hotel through the queen's gardens. A gardener once spied him, and told him to go around. He refused. The queen was told.

On 15th November 1899, information for the first newspaper ever produced at sea *The Transatlantic Times* was transmitted from the station by wireless telegraphy and printed on the U.S. Liner *St. Paul* when 36 miles distant.

On 3rd June 1898, Lord Kelvin sent from the Needles Wireless telegraph station the first radio telegram for which payment was made.

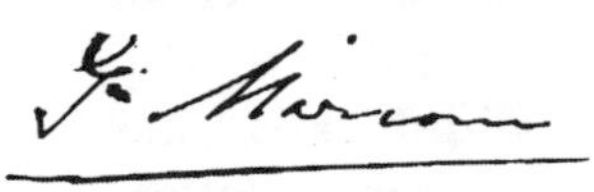

THE TRANSATLANTIC TIMES.

VOLUME I. NUMBER 1.

THE TRANSATLANTIC TIMES

Published on board the "*ST PAUL*," at Sea, *en route* for England, November 15th, 1899.

One Dollar per Copy in aid of the Seamen's Fund.

Mr. W W Bradfield, Editor in-Chief. Mr T Bowden, Assistant Editor. Miss J B Holman, Treasurer. Mr H H McClure, Managing Editor.

Through the courtesy of Mr G Marconi, the passengers on board the "St Paul," are accorded a rare privilege, that of receiving news several hours before landing. Mr Marconi and his assistants have arranged for work the apparatus used in reporting the Yacnt Race in New York, and are now receiving dispatches, from their station at the Needles. War news from South Africa and home messages from London and Paris are being received.

The most important dispatches are published on the opposite page. As all know, this is the first time that such a venture as this has been undertaken. A Newspaper published at Sea with Wireless Telegraph messages received and printed on a ship going twenty knots an hour!

This is the 52nd voyage eastward of the "St Paul." There are 375 passengers on board, counting the distinguished and extinguished

The days' runs have been as follows ;—

Nov. 9th	435
,, 10th	436
,, 11th	425
,, 12th	424
,, 13th	431
,, 14th	414
,, 15th	412

97 miles to Needles at 12 o'clock, Nov. 15th.

BULLETINS

1.50 p m. . . . . — . First Signal received, 66 miles from Needles

2-40 "Was that you "St. Paul"? 50 miles from Needles.

2-50 Hurrah! Welcome Home! Where are you?

3-30 40 miles, Ladysmith, Kimberley and Mafeking holding out well. No big battle. 15,000 men recently landed.

3-40 "At Ladysmith no more killed. Bombardment at Kimberley effected the destruction of ONE TIN POT. It was auctioned for £200 It is felt that period of anxiety and strain is over, and that our turn has come."

4.00 Sorry to say the U. S. A. Cruiser "Charleston" is lost. All hands saved

The thanks of the Editors are given to Captain Jamison, who grants us the privelege of this issue

The first ocean newspaper, autographed by Marconi on the date of publication, November 15, 1899. (Note Marconi signature top left)

"Who is he? What does he do?" she asked.

"Well, he's experimenting with electricity and wireless signals," they told her.

"Get another electrician," she commanded.

"Alas, your Majesty, England has no Marconi." She sent for him, heard his story, was impressed and (like his own queen) wished him success. This anecdote has two characters who typify, in some ways, the 19th century (her very name was given to that era) and the 20th century, with its fast communication.

Marconi's triumphs in early 1899 (the cross-channel link, Goodwin light-vessel message, etc.) were preludes to his first trip to America. He successfully reported the America's Cup races, but failed utterly to interest the U.S. Navy. On his return journey to England, he made ardent advances to a fellow passenger, Josephine Helman of Annapolis. This was destined to go the way of his other shipboard romances, but took three years to founder. On this trip news was collected from Poldhu in Morse, and the first ship's newspaper was concocted as the St. Paul approached England. The passengers got news about the South African war (see photos).

All this was a prelude to Marconi's Great Aim -- to span the Atlantic. The English site was on the long granite finger of Cornwall, pointing west. Several 200-foot-high masts in a ring 200 feet in diameter were erected. It took nearly a year to build this transmitting station, but it was all knocked down by a storm one night in September, 1901. The masts were scattered like jackstraws (see photos). Marconi impatiently replaced them by two 100-foot masts. They would have to do. In the meantime, he got the famous 7777 (four sevens) patent by which wireless signals could be isolated and clearly received (this was the syntonic tuning). The first choice for reception at the American end was Cape Cod. Only weeks after the Poldhu aerials had been blown down, another storm demolished the Cape Cod installations! Marconi transferred operations to Newfoundland. On November 26, 1901, Marconi, with Kemp and Paget, sailed for Nova Scotia.

Massive array of antennae at Poldhu, Cornwall, England, to conquer Atlantic. Erected 1901 summer. (Similar to ones erected in Cape Cod, U.S.A.)

Blown down 17 Sept. 1901 by storm. Cape Cod one similar fate on 26 Nov. (U.S. operations moved to Canada.)

Replacement by two masts only. From here signals sent —letter S— and received in Newfoundland 12 Dec. 1901; there balloon tried, but frequently blown away; so kite used successfully to receive signal.

Raising the kite at Signal Hill (apt name!) St. Johns Newfoundland weather too bad for balloons; masts already blown down. It will haul 400 feet of wire aloft and (hopefully) receive signals sent from Poldhu... Marconi on extreme left.

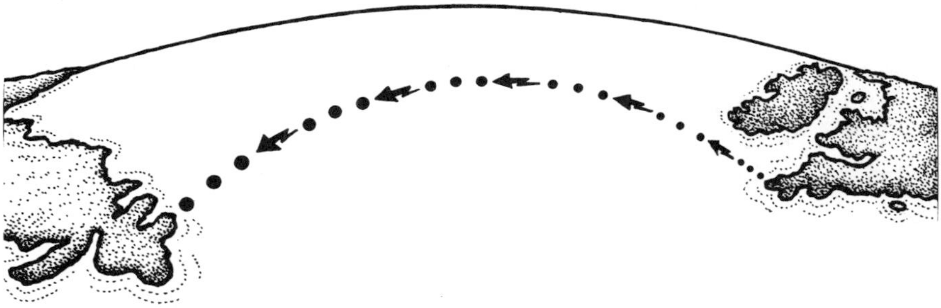

dit ... dit ... dit ... It came through at 12:30 p.m., Dec. 12, 1901. All the experts said it was impossible. Curvature of earth was "barrier."

Triumphant trio, (looking a little uncertain; this was taken earlier). George Kemp on left, I.P.W. Paget on right. Well-dressed Marconi was always a "dandy."

On arrival, they found the weather so bad and the time so short, they gave up the whole idea of masts. They used a balloon to raise an antenna. Even that was no good. They ended up flying a kite, to which were attached 400 feet of wire.

At midday on December 12, 1901, they were sitting at a table with a crude receiver (a few coils and condensers) attached to a coherer. All mathematicians and scientists were agreed that the curvature of the earth would stop the propagation of wireless waves. Just after 12:30 PM, click! click! click! (sharp but not loud) sounded in the earphone Marconi was wearing. "Can you hear anything, Mr. Kemp?" he said, handing the phone over. Kemp did. From Poldhu, 2,170 miles to the east, the signals were arriving. Then they stopped, probably because of the changing capacity of the antenna, buffeted by gale winds and floating up and down with the kite. At 1:10 PM -- click! click! click! again (still sharp and quiet). Next morning the New York Times announced in a dispatch from St. John's NF Dec. 14. "The most wonderful development of modern times...."

Unlike Alfonso's rifle shot -- the other great moment in the early history of radio, but known to very few people--all mankind was pleasurably excited, except for the Anglo-American Cable Company! They claimed a monopoly on radio-telegraphy in Newfoundland. They threatened Marconi with legal action unless he dismantled all his wireless gear. Cable company shares fell heavily on the Stock Exchange.

In the chorus of doubts and disbeliefs, the only comment which disturbed Marconi was the voice of his old friend and supporter, William Preece, who pointed out that Morse for the letter "S" -- dit-dit-dit -- is the most common phenomenon in natural atmospheric disturbances. However, the banquet for Marconi at the Waldorf-Astoria in New York, January 13, turned into an ovation. The word "MARCONI" in electric lights was stretched between two tablets marked, respectively, "POLDHU" and "ST. JOHNS." The lights flashed on and off in sequences of dit-dit-dit. Letters of congratulations came from Thomas Edison, Nikola Tesla, Alexander Graham Bell, and many, many others. Marconi took a ship back to England and, avoiding all fanfare, stayed with his mother for a month.

Then he sailed back to the U.S., going from Cherbourg on the American liner *Philadelphia*, with an entourage of his technical staff. The idea of this voyage was to blazon the truth as to just what wireless could do, how far it could reach. The records speak for themselves. From Poldhu messages were received at distances of 250, 464, 1027, 1065 and 2066 miles. Nighttime signals went 2,000 miles with no difficulty. The coherer had been the weakest link in the chain; Marconi had substituted for it a "magnetic detector."

One of the wireless operators has left us an account of what it was like in those days to send. The enormous Morse key was actually pounded. It was too heavy to send at more than 10 words a minute. Enormous sparks flew out, passengers and others were disturbed by the noise, and the operator became very tired after a few minutes of pounding. Contrast this with the way it was to be -- an operator would be using a bug key, sending 30 words a minute, operated by tiny wrist movements as he lolled back in his chair.

Marconi's roving eye for female beauty never rested long. He was smitten by, then engaged to, Inez Milholland. Except for her fantastic beauty, she represented nearly everything of which he disapproved. She was a fierce talker and a pioneer suffragette. The engagement was soon broken off.

During one visit to the U.S., he saw Thomas Edison at his home in Orange, New Jersey. They talked shop far into the night and forgot to eat. Marconi left Edison to return to London, on his third successive Atlantic crossing *without* a shipboard romance. On the rebound from the glamorous Inez Milholland, he had gone through the stages of being sulky then neutral then ready to approach women again. This time he really went overboard (figuratively speaking, not a shipboard romance). He fell desperately in love with the beautiful 19-year-old Beatrice O'Brien, daughter of the Irish baron Inchiquin. Her world was astronomically different from his. She was descended from Brian Boru, the warrior king of Ireland who was slain in his tent in 1014. Bea had seven sisters and six brothers, with a vast estate in Dromoland in County Clare. When they used to visit their English cousins, they used to hobnob with the royal family in a nearby estate.

By this time, Marconi had become famous and she couldn't help hearing about him, especially when she was visiting friends not far from Poole in Dorset, where Marconi was busy. He proposed

COHERERS -- the original Wire-less RECEIVERS

-- devised by Edouard Branly (French professor) in 1890.

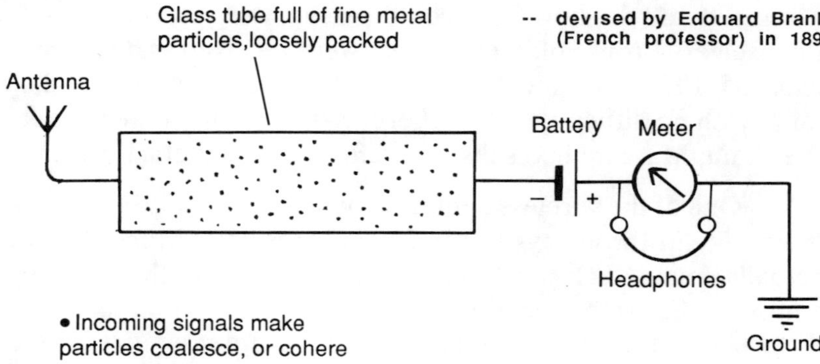

- Incoming signals make particles coalesce, or cohere
- this reduces their resistance
- current (from battery) can then and does, flow
  -deflecting meter
  -be heard on phones
  -make marks on paper

(Coherers didn't last long.They were unreliable and slow.)

1903 Poldhu. The Prince and Princess of Wales (to be King George V and Queen Mary in 7 years). Marconi, wearing "boater". Those enormous structures (to get height for antennas) weren't really necessary. Marconi thought they were, to get distance and power to cross the Atlantic. The magic and ease of short wave from small antennas hadn't yet been discovered.

to her but was turned down. Like the traditional suitor in a Victorian novel with a "broken heart," he went wandering around the Middle East on company business. Next year, a mutual friend got them together again, same place. Marconi proposed again, and this time he was accepted--on the condition that her sister Lila approved.

Bea's own objections were that she didn't love him -- she said so -- and that he was a foreigner, sometimes with outrageous behavior. Fortunately, he was not a practicing Roman Catholic; this would have made him, to the Irish Anglicans, the devil incarnate!

Anyway, despite misgivings, Bea consented (sister Lila had already okayed him).

"Marky," as he was called (Lord Inchiquin especially couldn't cope with all the g's and l's in "Guglielmo") wed Bea at St. George's in Hanover Square -- then, as now, the most fashionable church in London. The Italian ambassador was present. Presents poured in -- including a sealskin coat and silver samovar from Popov. MARCONI WEDS DAUGHTER OF IRISH PEER went the headlines. The public swarmed to the church. Guglielmo and Beatrice honeymooned in Dromoland -- for one week (businessman Marconi couldn't afford more time). Bea's statement, "I don't love him," hadn't stopped her from marrying him, and by the end of the honeymoon she was deeply in love with him.

Back in London, they stayed in a small hotel near Marconi's office. Full of Italian jealousy, Marconi used to require his wife to tell him how long she'd be gone, where she was going, what streets she was going to -- everything. She was flattered to be so cherished, but felt confined.

In his wireless world, Marconi felt his gear was out of date. He wanted to get a more powerful station built in Clifden, Ireland (not far from Dromoland). To establish an American station to get the signals, he set out again for America. He spent most of his time on the voyage in the wireless office. Pretty, friendly Bea -- now the wife of a famous man -- enjoyed sharing her excitement with fellow passengers. To keep her by him, jealous Marconi instructed her in the Morse code! This was a bore to her.

She perked up when they got to New York with its round of parties and dinners; even more so with President Theodore Roosevelt and his daughter at Oyster Bay, but that period was brief.

Marconi took her on to Glace Bay in Canada, where he was establishing a wireless station. The place was as forbidding and cold as it sounds. Beatrice tried to walk to the nearest town, but didn't realize how far it was. She turned back. Marconi -- and everyone -- had been frantic with fears about her. She stumbled in about 8 that night. Marconi scolded her severely. She deserved it, and didn't forget.

Back in Poldhu, life was equally dreary, so she stayed in London. Marconi came up to her as often as he could, but the trip took 11 hours, and he had little time to spare. A daughter was born to them in February, 1906.

The radio-telegraphy between Glace Bay in Canada and Clifden in Ireland was now operating very well. In October, 1907, 10,000 words were exchanged via radio telegrams. (It was five years, ten months since the first message had been sent over the ocean.)

Marconi was in America again when his wife bore a second daughter, Degna.

On January 23, 1909, the liner *Republic* was rammed by the Italian ship *Florida*. Her wireless cabin was crumpled, but the instruments therein and the Marconi operator were okay. Jack Binns, using the emergency accumulators, sent signals to the nearby Siasconset coast station. They alerted other ships, who headed for the stricken liner. The *Baltic* was the first to arrive. Some passengers were transferred to her, some to the *Florida*. Seventeen hundred people were saved, largely because of this new device, wireless.

In New York, Liverpool and in his home town of Peterborough, Jack Binns became a new kind of hero -- a wireless operator. He was mobbed, feted, given awards, etc. He deprecated all this, saying he'd only done his job.

When Marconi went to America in 1909, Bea stayed home with Degna, having found out that she was pregnant again. When

Marconi returned, Bea had planned a gay escapade. Instead of waiting to meet him at the dockside, she crossed to Ireland and boarded a small boat in Cork, to meet Guglielmo as he arrived. He was angry when she popped out of the sea like a mermaid. They had a hot scene. It was the first of many violent storms which eventually (but not until after World War I) would lead to their divorce.

In the meantime, however, they were on good terms again in December, 1909, when it was decided that she should accompany him to Stockholm where he was to receive the Nobel Prize for physics. This turned out to be a joint award -- to him and to Ferdinand Braun of Germany. Braun was as astonished as anyone else. He thought, too, that the award should have gone to Marconi alone. They parted as good friends.

Marconi went wandering, alone, all over the world. While on a trip to South America, he kept good contact with Clifden -- once at a distance of 6,725 miles. He wanted Bea to be in Italy for the arrival of their next child, which he felt sure would be a boy this time. He was right. Julio Giovanni Victoria Marconi was born on May 2, 1910. Marconi was in the Atlantic somewhere, on a ship unknown to Beatrice. How to inform him? No sweat. She sent a telegram addressed to MARCONI ATLANTIC! It was relayed from ship to ship until he got it!

A happy summer was spent at Grifoni, but the idyll did not last. When they went back to England, the tensions between Guglielmo and Bea were serious. Marconi tried to keep up with social life in London, but his workload was too heavy. He got nervous and irritable. He was, as always, jealous. They got near the breaking point. His friends and even his mother (surprisingly taking the side of Bea) persuaded them to make up, and he took Bea on an automobile trip through Italy. Unfortunately, an accident near Pisa resulted in the loss of Marconi's right eye. His artificial right eye was made with such skill that many never knew he was sightless there.

Tension between him and Bea continued, but they didn't let this affect the lives of their children or didn't even let them know what was going on.

They were invited to be passengers on the maiden voyage of the *Titanic*, but fortunately (though they'd actually booked cabins) this was not to be. Marconi had switched his passage to the *Lusitania* because he had a mountain of paperwork to clear, his secretary was a hopeless sailor (he'd be sick from shore to shore) and the ship could provide him with a secretary. They left three days before the *Titanic*, on which Bea would have been a passenger but she had to postpone the trip because Julio came down with a baby fever.

On the morning of April 10, 1912, Bea and her kids saw the *Titanic* from their beachside home after it left the docks at Southampton. The ship was huge and resplendent in the spring sunlight. They waved and cheered. Dozens of handkerchiefs waved back at them. They, with Marconi, should have been among the hankie-wavers.*

The *Titanic* disaster and Marconi's part in it (in the sense of the wireless gear) is covered elsewhere. He was still getting praise (for the 712 who were saved by wireless) when the infamous Marconi scandal broke.

He'd made plans to link the British Empire by wireless stations. There were rumors of a "Marconi ring," company promotion, etc. -- all quite false. Marconi was innocent, and at this distance in time it seems like a teacup storm. Sportsmanlike apologies were made, and King George V capped them all by making Marconi an honorary Knight of the Grand Cross of the Royal Victorian Order. He could have called himself thenceforth Sir Guglielmo, but he never did.

On May 29, 1914, the *Empress of Ireland* collided with a Norwegian freighter at Father Point on the St. Lawrence River. Damage was severe and the ship was soon sinking. Her wireless operator, Ron Ferguson, sent a call for help and was heard by the wireless station at Father Point. In just eight minutes, the power failed and the ship sank, but two ships were coming to the rescue. Ferguson was in the icy water for a quarter of an hour before being picked up. He was one of the 444 who survived. 1023 died. There'd have been many more, without the SOS.

**See what they saw on page 275.*

In 1915, just before Italy entered WWI, Marconi was in America. He came back to Italy and volunteered for the armed forces. He was sent to inspect wireless units at the front. Back in England, Beatrice gave birth to another baby girl on April 10, 1916. They tried to join Marconi in Italy, coming via Switzerland. They arrived disheveled, like refugees ("All you need is a parrot," Marconi remarked). He established his family at the Excelsior Hotel in Rome, and took one room there as a laboratory where he spent long hours experimenting. When he was transferred to the Navy, he built a huge antenna on one ship that could receive signals from all over the world. Back in Rome, he built himself a receiver through which he could get important news before anyone else. On November 5, 1918, his daughter Degna had just come in to visit him. He had earphones on. He bent over to whisper in her ear, "The Kaiser has abdicated." At that moment, Guglielmo and his 10-year-old daughter were the only two people in Rome who knew this.

The war years had been a sort of reprieve for the Marconi marriage, which had been heading for catastrophe. After four years, peace came to the world, but the Marconis lost theirs. He'd often been fooling around with other women, but now he was quite brazen about it. However, he had a new distraction, away from all females, a 220-foot yacht originally built in Scotland for the Arch-duchess of Austria. He began to use it as a floating laboratory, naming it *Elettra* ("spark") and he began to experiment on her endlessly, everywhere. He was busy outfitting the ship when his mother died. She was well over 80. He couldn't find the time to go to her funeral. When at last the yacht was ready for cruising, Marconi telegraphed Beatrice in England to join him on the maiden voyage.

There was little chance of patching up their differences, especially when she found that her latest adversary (with *her* husband!) was on board! Nevertheless, the cruise continued -- to Spain, Gibraltar, Morocco, and back to Italy. Marconi was busy in the radio cabin most of the time. Strained relations between Guglielmo and Beatrice continued for another three years. In the meantime, he did a lot of radio work, especially in connection with "beam" radio on shortwave, which was supplanting the old long-wave system.

Suddenly, Beatrice asked him for a divorce. He didn't really want it. It had never occurred to him that after all his philanderings Beatrice might say "enough's enough." If he'd refused, and made real efforts to reform, she *might* have forgiven him. He was always to be sorry that he didn't do this. Bea conferred with her friend, the queen of Italy, who said, "A man like Marconi should never marry." The fate of their children was settled amicably. They were to spend three months in Italy with Beatrice and their summer holidays with Guglielmo.

Bea remarried. Many, many letters were going to come to her from Marconi in years to come, every single one of the starting, "Dearest Bea," and ending with "Yours affectionately, Guglielmo." One could say that each carried a torch for the other almost to the end of their days.

He then took up with Cristina Bezzi-Scali, a quiet girl with fair hair and blue eyes, the very opposite of the gay, sophisticated women he'd been cultivating. Her strict Roman Catholic family would *never* permit their daughter to marry a divorcee. The only way the Catholic Church could give its blessing to Marconi's second marriage was to declare his first annulled. To do this, Marconi had to prove that there had been mental reservations either on his part or on Beatrice's about their projected marriage while they were still courting. That was done. It took some time. Tongue-in-cheek attitudes, bending over backward in involved arguments -- turning a blind eye to the evident love Bea and Guglielmo had had for so long and were continuing even now--it was all navigated past triumphantly. His first marriage was dissolved. Guglielmo and Cristina were wed on June 12, 1927, in a civil ceremony. They sailed for their honeymoon on Marconi's 85th Atlantic crossing. The bouncy mayor of New York, Jimmy Walker, welcomed them to the "biggest Italian city in the world." They had a good time. When they got back to London, Marconi started getting sharp pains in his chest. It was diagnosed as a severe case of angina pectoris. After two weeks in a nursing home, they moved back to Italy.

Cristina, sweet as she seemed, wanted to be the center of her man's life and all strong links with the past were swept away. Captain Lauro, the autocratic master of *Elettra*, was dismissed. Marconi's "indispensable" secretary of 20 years came no more. Marconi himself reverted to strict Catholicism. A little girl was born to them in 1930 -- she was called Elettra. Degna and Giulio were no longer invited to cruise with their father and Cristina. Their mother's allowance was also reduced. Beatrice summoned her lawyer: this was the signal for open war.

In 1935, Marconi wrote a new will, leaving everything to Cristina and their daughter, Elettra. This put Bea's children "beyond the pale." The estrangement between Marconi and his three elder children was almost absolute.

In his radio world, however, he made progress as never before. The severe tensions between the Marconi Company and the cable companies (dating back, you may remember, to that initial "S" across the Atlantic) was resolved by a marriage of the giants. In the merger of Cable and Wireless, the Marconi Company was swallowed up.

An Italian expedition to the North Pole in 1928 had taken, in addition to the ordinary broadcasting equipment, a shortwave transmitter, built by the Marconi Company. When the ship crashed, 10 of the men were thrown out on the ice and six were carried to their deaths. One of the saved was the radio officer, Biaggi. Using the shortwave transmitter, he called for rescue. A Swedish airplane spotted the red tent and rescued the survivors.

Next year, Admiral Byrd got to the North Pole by airplane. Listeners everywhere followed his flight by radio. Proponents of wireless and aviation were given a great boost by Italo Balba in 1933, when he led the entire Italian air fleet in maneuvers by instruments and shortwave in Chicago.

Marconi was an avowed political innocent, and had no idea what Mussolini was up to. In the end, he went along with him because his king supported Mussolini and his Fascists. Mussolini wanted the glory that Marconi's name would give to his party. He used to visit the *Elettra* and showed great interest in the development of wireless.

Marconi, wanting the country of his adoption, England, to understand the country of his birth, was going to broadcast an explanation of Mussolini's invasion of Ethiopia. The BBC refused him.

In 1933, Poldhu sent its last message. Ultra-short (10 to 1 meter) waves and even quasi-optical waves (under 1 meter) were taking over. Poldhu had been the last proponent of longwaves (up to 30,000 meters).

In 1931 there were commemorations of the original Atlantic spanning. This was the last public appearance of Kemp (now white-haired). He died just over a year later. Marconi had another six years to live, and many sea miles yet to go on his beloved *Elettra*. He made a memorable trip around the world with Cristina, visiting his old friend and one-time employee David Sarnoff in Chicago, where he tapped out "S" in Morse. These three dots circled the globe via New York, London, Rome, Manila, Honolulu, and back to Chicago in 3 minutes and 25 seconds. He met President Franklin Roosevelt, who remembered their previous encounter in 1917, when Roosevelt was Naval Secretary. He went on to California, stopping at the University of Notre Dame, where he was given an honorary degree. On to the Grand Canyon, Hollywood, San Francisco (where he was made an honorary citizen), and then Yokohama. They were reunited with their daughter in the spring of 1934, and then sailed on the yacht *Elettra* back to Italy.

After this, the heart attacks became more frequent and severe. On April 25, 1937, Marconi's 62nd birthday, when he was in the Hotel Splendide in London, he learned that his brother Alfonso was ill; soon afterwards, Alfonso died of the same heart disease that had killed his mother, and would soon kill Marconi. Alfonso was buried beside his mother.

The one which finished Marconi came only three months later, at 3:45 AM on July 20, 1937. A worldwide silence was declared by all radio stations everywhere, to honor the man who'd made it all possible.

Let's return to Marconi House in London in 1931, where Marconi, Kemp and Paget were celebrating. Marconi said, "My thoughts go back to 30 years ago, when I was standing on the top of a hill in bitterly cold Newfoundland, wondering if I should ever be able to hear the letter 'S' coming from England." He, Kemp and Paget recalled their youthful camaraderie. There was a cold wind outside. Kemp used to keep a saucepan of cocoa on the back of the little stove to warm Marconi as he came in.

Let's transplant ourselves in time and space, back to that moment when Marconi had just come in from the bitter cold outside, anticipating Kemp's hot cocoa. His hands are nearly frozen from trying to handle the cable holding the kite. Finally, he sits down at the table and listens for what he is anxiously hoping to hear ... and *does* hear ... click! click! click! ....

Sir Ambrose Fleming 1849-1945 Inventor of the Diode

He helped introduce wireless telegraphy, electric lighting and the telephone into his homeland of Britain. His major contribution to the development of wireless communication was the construction of the diode, a two-electrode vacuum tube. Thomas Edison in the 1880s had observed a "cloud" of electrons passing from a hot filament to a cold metallic plate inside an evacuated tube. This was to be known as the "Edison effect," and was important because it was the basis of pretty well everything connected with radio tubes. Actually, Edison (one of the greatest inventors ever) gets more credit than he deserves. He only noted the cloud of electrons. He never did a damned thing about it!

The Edison effect seemed to imply that an electrical current was passing across the space. In 1904, Fleming proved it was. He used the two-element vacuum tube to change the small AC current into a big DC current. This DC signal could then be picked up by an electrical meter or a telephone receiver. By blocking off the negative side of the AC current and turning it into DC, the tube was acting like a valve. (That's what the British still call it, a valve.) It was a great step forward in the accuracy and speed of wireless reception. Fleming later went to work for Marconi. He took life a lot more easily than his high-strung boss, and lived to be 94.

"MR. RADIO"
Dr. Lee De Forest
1873-1961

It's one thing to send dots and dashes from place to place through the ether, and quite another to send the human voice. Both are effective means of communication, but the Morse code is essentially a different alphabet -- sometimes a different language -- from the one people normally use. The next great development in radio was the transmission of the human voice, and for this the credit goes to Dr. Lee De Forest.

His enormous gift to mankind was the invention of the triode, or, as he called it, the audion. When, in 1906, he inserted a grid (a zigzag platinum wire) into Fleming's diode, it meant that there was now *control* by external means. That little bent piece of wire gave birth to the electronic age. Ultimately, it meant that spacecraft could send readable signals from millions of miles out in space. It was one of the most valuable patents ever issued by the U.S. Patent Office.

De Forest had over 300 patents, but to appreciate this one you'd have to measure: all the laughter, tears, oohs and ahhhs ever experienced in a movie theater, or in front of a TV, or in listening to a voice or music radio broadcast, plus all the information ever carried on long-distance telephones, plus all the electronic circuitry ever devised. All these resulted from the invention of that precious audion.

De Forest lacked Marconi's ruthlessness in business. He was a gentle, almost saintly, man, and he lost three fortunes by letting others swindle him blind.

* * * * *

He was born in Council Bluffs, Iowa in 1873, a year before Marconi. He grew up in Alabama where his father was president of a small college. As a student at Yale, he kept pretty much to himself because he was constantly inventing things. He threw himself into the study of radiotelegraphy and electricity, concentrating on the problem of making a more efficient coherer which could receive signals across land distances as easily as across water distances. His Ph.D. dissertation at Yale (1896) was entitled, "Reflection of Short Hertzian Waves from the Ends of Parallel Wires."

We'll list his ups and downs henceforth chronologically.

1901. He perfected the electrolytic receiver, a vast improvement on the coherer. The basic principle was this: some chemical solutions conduct electricity better when an electrical wave is present. When both electrodes of a radio receiver are dipped in such a solution, a current will be generated between them, depending on the amount of conductivity. De Forest called his device a "sponder." It was superior to all other receivers in speed and accuracy.

1902. He became involved with a stock promoter, who organized the De Forest Wireless Telegraph Company. To use the lingo of about 20 years later, he "was taken for a ride." This was the first of several catastrophes.

1904. The counterbalancing upswing came at the St. Louis World's Fair. His sponder was awarded a gold medal. The new receiver had already been pressed into service in the Russo-Japanese War by New York and London reporters. De Forest's triumph at St. Louis was complete: with the sponder and all the other wireless/radio inventions he demonstrated, he took ALL the prizes in the field! (Incidentally, it was about this time that the word "wireless" began to be replaced by "radio." Strictly speaking, a completely wire-free gadget is impossible. All the filaments, grids and connections in a radio tube are wires! It's the distance between transmitting station and receiving station which is "wireless." It is historically fitting that Marconi should be known as Mr. Wireless and De Forest as Mr. Radio.)

1905. Downswing! The directors of his company were, without his knowledge, involved in unscrupulous practices which eventually sent them to trial and prison. De Forest was also charged but acquitted. First fortune was lost.

1906. Big upswing. He added a new wrinkle to his momentous invention of the triode. Fleming's tube, the diode, had *detected* waves. Thus,

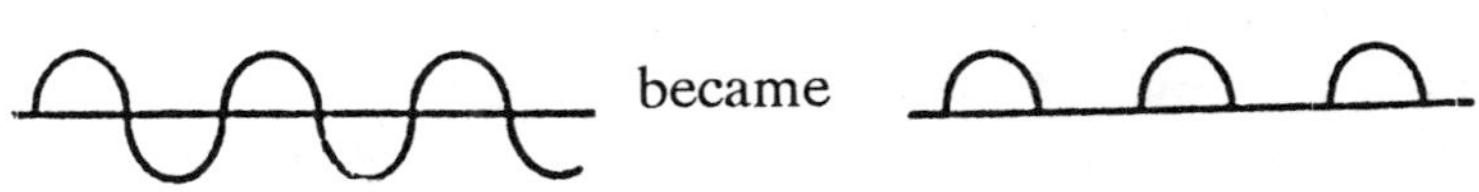

De Forest's triode, with its extra grid, made amplification possible. Thus,

 became 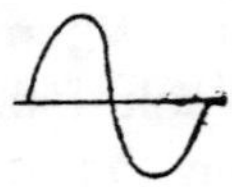 (or as loud as desired)

Also, and more importantly, modulation was possible. This opened limitless horizons. Voice and music could ride piggyback on the original carrier wave. 

1907. In this year he started the first arc-phone broadcasting system. This he sold to the U.S. Navy for use in their famous trip around the world as the Great White Fleet. Each ship was able to speak directly to the others.

1909. His greatest invention, the triode tube, came into its most glorious use. With all the air pumped out of the tube, forming a vacuum, De Forest's triode performed fabulously better than it had before.

1910. A New York Metropolitan Opera broadcast became the first-ever over radio-phone. The voice of Enrico Caruso was heard by those with receiver sets -- including ships at sea.

1912. The audion was advanced to a far-reaching amplifier when Western Electric supplied De Forest with a better vacuum tube. Endlessly fiddling with amplification, he made it possible for the Navy, in Arlington, Virginia, to arrange 500 of his tubes in series (linked one to another, bolstering the power) and to transmit the human voice to Paris, then on to Honolulu.

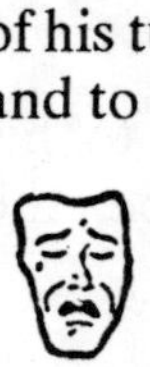

1913. "A patent is just an invitation to a lawsuit," said De Forest. It turned out that Major Edwin Armstrong made the same discoveries simultaneously with De Forest. A long, costly lawsuit ensued. De Forest won this one ... eventually. In another sense they both lost. De Forest was penniless for quite a while, and poor Armstrong was only beginning his tussles with others. Finally, worn out, he committed suicide in 1954.

1914. De Forest was involved in one of several lawsuits with Marconi; this over the Fleming diode patent (Marconi won).

1915. He saw his audion being used when the first transcontinental phone conversation took place between Washington, D.C. and San Francisco. Edison and Bell profited (publicity and money) but De Forest's audion wasn't even mentioned in the coverage. His pride hurt, he fought back, printing many statements about how the De Forest amplifier had made the achievement possible.

1916. He organized the first real broadcast station, KDKA in New York City, using a 125-watt tube, the biggest ever. He was even the first disc jockey.

During World War I, his audion tube fascinated amateur radio buffs, who were scouring radio shops for galena crystals, headsets, etc., to receive broadcasts. "How much distance did you get last night?" was a standard morning greeting. (Both Marconi and De Forest were amazed at the way the public had taken to wireless/radio.)

1917. He sold all his audion rights to AT&T for $250,000 and used the money to buy more experimental machinery.

1920. He got diddled out of another fortune by signing over one of his precious patents to an attorney who had expressed an interest in his work. He had been suspicious that the guy represented a big corporation, but had been assured it was not so. He let his patent go for $50,000 when he could easily have gotten $500,000.

1923. He became interested in the idea of putting a sound track on film. His "phonofilm" did just that. He tried peddling it in Hollywood, but no one was interested. Who wanted sound on moving *pictures*? His failure was probably his own fault. He was the world's worst salesman. However, he did manage eventually to sell his patent for a small sum.

1927. When Warner Brothers produced the first movie with synchronized sound *(The Jazz Singer),* it was an instant success. It should have been a bonanza for De Forest, but he would have had to war against a phalanx of attorneys and Western Electric. If his name had been Marconi, he'd have hacked it against them all, but De Forest was too kindly. It was his last big business failure.

1931. For the first time in years, he was *not* involved in patent lawsuits. Those still outstanding -- for high-vacuum tubes, feedback circuits, regenerative patents, etc. -- were all settled in his favor. Even the animosity seemed to have left everyone.

1935. This year saw his last important invention -- for TV scanning. Unfortunately, the world was not quite ready for TV and he lost interest.

1937. Never good at making money, in this year he declared bankruptcy. His patented inventions enabled him to live comfortably and to continue tinkering, which was his great love in the first place.

1943. To celebrate his 70th birthday, he climbed Mt. Whitney.

.... So far, women haven't been mentioned. Like Marconi, his sex life was mostly a mess, but his final marriage was a success. Of marriage failures, he had three. (He made and lost three fortunes, wed and lost three women.) The fourth took. It was to the beautiful movie stunt actress Maria Mosquite. She can be classed as the greatest asset of his long and colorful life. They were very devoted to each other. His last words to her were, "I love you more than words can tell."

He died June 10, 1961. He was 87.

He had been experimentally active in electronics almost to the end; his last patent had been granted four years earlier. Without Marconi's drive -- which naturally involved debilitating tensions -- he outlived his contemporary by 24 years.

SOME CONVENTIONAL DIAGRAMS OF BASIC RADIO FEATURES

A regular, even wave; "CW" (continuous wave).

A "damped" wave from *Spark-Transmitter.* After the first big effort, or push, it dies down gradually.

CW, showing amplitude (strength).
Above the line is +.
Below the line is −.

The negative half of the wave has been cut off, in Fleming's invention.

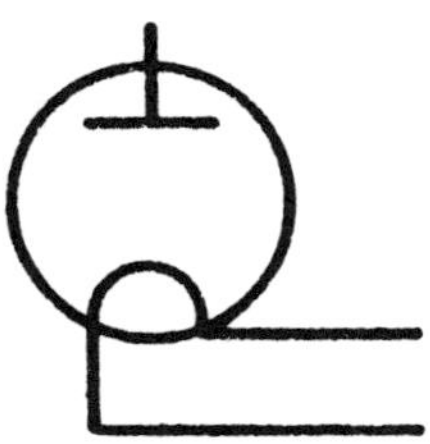

Fleming's "diode" (Am: "tube"; Brit: "valve"). Electrons go from filament (below) to anode or plate on top.

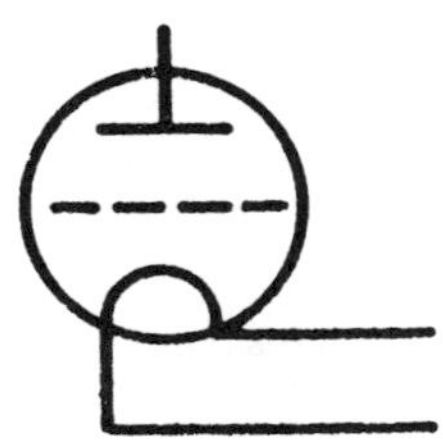

DeForest's "open Sesame" to nearly everything! The "audion" or "triode". The center GRID gives *control*.

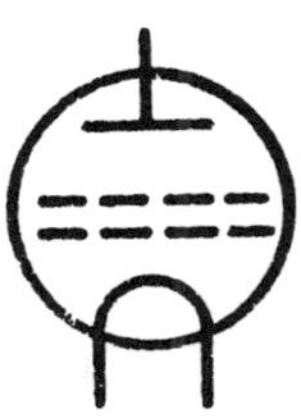

A tetrode (4 elements/filament, 2 grids & anode).

Pentode (5 elements), and so on. More grids — more sophistication/control, etc.

The Inventors Responsible For Radio

Item	Date	Inventor	Nationality	Flag
Leyden jar	1746	Van Musschenbroek	Dutch	
Induction coil	1842	Masso-Breguet	French	
Spark Gap	1887	Righi	Italian	
Oscillator (all above)	1887	Herz	German	
Coherer detector	1890	Branly	French	
Elevated wire aerial	1895	Popov	Russian	
Elevated wire aerial	1895	Marconi	Italian	
Ground connection	1895	Marconi	Italian	
HF Transformer	1895	Marconi	Italian	
Diode/tube/valve	**1904**	**Fleming**	**English**	
Audion (radio tube with grid)	1906	De Forest	American	
Speech/music broadcast	1906	Fessenden	American	

MARCONI WAS THE FIRST TO ASSEMBLE VARIOUS COMPONENTS AND MAKE THEM *WORK*. He said "The idea was so elementary, it was hard for me to believe that it had not been already tried by someone."

MR. ELECTRICITY
Nikola Tesla
1856-1943

He has been called "the man who invented the 20th century," but most people have never even heard of him. One of his inventions was the standard AC motor, which runs everything from kitchen refrigerators to all the massive machines in industry. Most people who read about him become his enthusiastic fans.

Nikola Tesla was born in 1856 in what is now Yugoslavia. At age 19 he moved to Graz, Austria, and received a sound education in math, physics and mechanics. Later he became preoccupied with the problem of wasted power in electric (DC) motors. Was it possible to stop the brushes from sparking and frittering away energy uselessly?

One day Tesla was walking with a friend in a Budapest park, spouting poetry from *Faust*, when the solution suddenly occurred to him. Excitedly, he demonstrated it to his friend by drawing diagrams with a stick on the dusty ground. It was basic, simple: Do away with the brushes! Scrap the commutator! Instead of moving an iron rotor through a magnetic field, the magnetic field itself should rotate, powered by AC, not DC.

This was the principle of the polyphase inductive motor--the most efficient use of AC and the beginning of Tesla's marriage (no better word; he never wed a woman) to alternating current. He couldn't convince his superiors at the telephone company, though he was chief electrician, so he went to Paris and joined the Edison Corporation. He still got cold, negative reactions to his ideas. He immigrated to America and approached Thomas Edison himself; still no go. Edison was firmly attached to DC, could only see the dangers and uncertainties of AC. He gave Tesla a job designing DC components, which lasted about a year, until there was a disagreement over wages due to Tesla. They remained divided on the AC/DC issue. Thereafter, Tesla was furious at even the mention of Edison's name.

He established his own company and developed an arc lamp to light streets and factories. Yet, just as successful manufacture was starting, he was "eased out of" the company by jealous partners. He was so poor that for a while he had to earn money by digging ditches.

In 1887 he again founded a company in which he was determined to develop his AC motor. Within two years he had received 12 patents and had impressed the American Institute of Electrical Engineering. He sold his patent for a polyphase inductive motor to George Westinghouse. He also joined the Westinghouse Company, but resigned after a year, itchy (as usual) for independence and after falling out (as usual) with his superiors over technical matters. He became a U.S. citizen in 1889.

He may be almost unknown today, but back in the 1890s he was making one hell of a stir in the scientific and social worlds. His lectures brought him worldwide scientific fame. As a social celebrity he fascinated everyone by his dramatic demonstrations; for example, he'd wave an apparently empty tube about, then light would suddenly appear inside it from a concealed Tesla coil. Showbiz like this, coming from a tall, handsome, hatchet-faced man, attracted women enormously, but he was . . . well, not exactly a misogynist; he didn't react to women at all! When no less a feminine persuader than Sarah Bernhardt tried the stereotyped ploy of dropping a handkerchief near him, he picked it up and returned it without a word. As mentioned, he never married. After a bout with illness, he withdrew from social life completely and concentrated on what he believed was his true destiny, to discover new scientific principles.

In New York his flamboyant manner and eccentric habits made him an object of intense interest. He slept about two hours per night. He aimed to be the best-dressed man on Fifth Avenue, with gray suede gloves, black derby, cane, etc. He discarded each pair of gloves after a week, and handkerchiefs after one use only, because of his germ phobia. He dined nightly at Del Monico's, then at the Waldorf-Astoria, demanding that no one else use his table even when he wasn't there. He required a whole stack of freshly laundered napkins at every meal; with them he'd carefully wipe each utensil and dish, then drop each napkin to the floor.

Despite his peculiarities, his scientific achievements reached their apogee during this period. His polyphase AC system proved to be a complete success when it illuminated the buildings and grounds of the Chicago World's Fair in 1893. He then went on (with George Westinghouse supporting him this time) to his most memorable success -- the harnessing of all the power pouring over Niagara Falls.

This was the world's first hydroelectric generating plant, distributing power all over upper New York State and even farther. Long-distance power transmission was now a fact. Edison couldn't do it with his DC system; power could only be sent a few blocks from its originating plant.

At the height of his success, let's step back from Nikola Tesla and make comparisons with Guglielmo Marconi, who at that time was playing around hopefully with condensers, spark coils, etc., on the top floor of the Villa Grifoni. Both had been inspired by the work of Heinrich Hertz, who had successfully confirmed the existence of electromagnetic waves. Both wanted to use this wireless medium and power. Marconi's aim was, and remained, long-distance *communication*. Tesla's was the *transfer of power.* The successful manipulation of Hertzian waves, using electricity, was their common tool.

In 1897 Tesla demonstrated his own wireless communication system over 25 miles, but his methods were not well conceived and radio developed along quite different lines. He rightly saw that such elements as an antenna and a resonant-tuned circuit would be needed. He is also credited with something else in radio history: while Hertz discovered medium waves and Lodge found short waves, Tesla originated long waves.

In 1898 he built a model of a radio-controlled ship and patented it; this was the ancestor of all remotely controlled craft (in air or on sea). He predicted, 20 years before commercial broadcasting, that radio would "prove very efficient in enlightening the masses, particularly in still-uncivilized countries and less accessible regions." He said it would involve "the employment of a number of plants, all of which are capable of transmitting signals to the uttermost parts of the earth." When Marconi sent signals across the Atlantic in 1901, Tesla's telegram of congratulations was there, alongside that of his old adversary, Edison.

In Colorado Springs he constructed a massive machine to build a potential of *100 million volts* at 300 khz. It rushed upward, hit a ball on the top and exploded into huge flashes of lightning more than 100 feet long. Crashing thunder could be heard for 13 miles. Tesla once blacked out the entire city by drawing too much current from the local generating station, causing it to collapse. In 1902 he built a gigantic tower on Long Island, topped by a huge copper ball. This was part of a worldwide power-transmission scheme, which failed for lack of money.

Over the next 40 years, he grew poorer and increasingly eccentric. He refused to be a corecipient of the Nobel Prize for physics with Edison in 1912 (though he could well have used the money). When Edison sent him an olive branch in 1917 by offering him the Edison Medal of the American Institute of Electrical Engineering, he was persuaded to accept it, but reluctantly.

He continued to toss off extraordinary ideas, some feasible, some ridiculous. He'd have been completely destitute if he hadn't already become a national hero in Yugoslavia; he was almost wholly dependent on the yearly gift of $7,200 from the Yugoslavian government. Ignoring his American citizenship, they claimed him as their son.

His personality certainly handicapped his undoubted genius. He wouldn't cooperate or adapt. He had no tact in dealing with large corporations. He insisted on complete independence and withdrew from human society. He spent hours feeding pigeons in New York parks and harboring them in his hotel room.

He died in his sleep in his New York hotel room in 1943. He was 86. His remains were cremated at Ferncliff Cemetery, Ardsley, New York. He merited the praise given in the New York Times obituary: "all mankind is his debtor."

PURE INDIAN TEA

BRITAIN'S BEST BEVERAGE.

Weekly Dispatch

109TH YEAR

SUNDAY SPECIAL EDITION.

ESMOLIN

NEURITIS

No. 5,677.] SUNDAY, JULY 31, 1910. [ONE PENNY.

CRIPPEN'S LIFE AT SEA DESCRIBED BY 'WIRELESS.'

DETECTIVE INSPR DEW

MRS CRIPPEN — (MISS BELLE ELMORE)

H.H. Crippen.

THE DOCTOR'S FIRST WIFE MRS CHARLOTTE CRIPPEN

CAPT KENDALL OF THE MONTROSE

MISS LE NEVE

CRIPPEN'S LIFE AT SEA DAY BY DAY.

DESCRIBED BY CAPTAIN KENDALL OF THE MONTROSE.

FULL EXCLUSIVE "WIRELESS" TO "THE DAILY MAIL."

THE BOOKS CRIPPEN READS, AND THE SONG HE LIKED.

WHY HE FAILED TO ANSWER THE NAME "ROBINSON."

"ALL WALKED INTO THE SHOP."

WHOSE REWARD?

Canadian Police Squabbling:

NOTICE.

A Special Edition of *The Weekly Dispatch* will be published this afternoon, giving an account of the arrest of Crippen and Miss Le Neve.

Special arrangements have been made for having the details rapidly cabled by our special correspondent from Father Point, where the liner Montrose will be boarded by the Canadian police officers and Inspector Dew, and our special edition will be published at the earliest moment after the receipt of the news.

WIRELESS AIDS MAN-HUNT

Everybody on two continents knew about Crippen and Le Neve by wireless messages to newspapers; on board the *Montrose* only the captain and wireless operator knew what was happening.

Dr. Crippen had killed his wife in London and had buried her chopped-up body under the floor of his coal cellar. Panicking after a very intensive interrogation by Inspector Drew of Scotland Yard, he went over to Holland to board the Canadian Pacific Railroad liner *Montrose* -- sailing for Canada on July 10, 1910. With him was his mistress-typist Ethel Le Neve. He'd disguised himself a little, and Ethel was to be his "son"; she was outfitted in men's clothes and her hair cut short.

Meanwhile, back in London, Inspector Drew had searched Crippen's flat, found the grisly remains of his wife, and had alerted Interpol, worldwide, to help track him down.

Captain Kendall, master of the *Montrose*, had been second mate of the *Lake Champlain* nine years earlier (1901) when this British ship had been the very first to be equipped with wireless, so he was well aware of its value. In his present position he was also aware of other things. He knew about Drew's search for Crippen and Le Neve; he had seen pictures of them in the newspapers. He was also very suspicious about two of his passengers who called themselves Mr. Robinson and son.

His suspicions sharpened. Mr. Robinson's son seemed very effeminate. "His" clothes were ill-fitting. "He" squeezed "his" father's hand affectionately. Then a little event swept the last doubt from Captain Kendall's mind.

Mr. Robinson tossed a magazine over to his "son," who spread "his" legs to catch it instead of bringing them together. This would make sense if "he" was a "she" who normally wore skirts. A woman would open her legs slightly to make a sort of catching-basin.

Captain Kendall immediately went up to the office of his wireless operator, comparing its appearance with that of the original radio "shack" on the *Lake Champlain*. Then it had been literally a wooden shack (probably the origin of the term) on the side of a funnel. Now, on the *Montrose*, its knobs, wheels and gleaming paraphernalia gave an atmosphere of efficiency.

"Sparks," said Captain Kendall, "all telegrams are, as you well know, a confidential matter between you and me; this one, particularly so."

This picture of Dr. Crippen and his mistress Ethel Le Neve (on left, disguised as a man) was taken by Captain Kendall as they strode the deck of the *Montrose.* Detective Inspector Drew (of Scotland Yard) was on another ship, overtaking them and . . .

. . . arrested them in Canada. Here, Inspector Drew is escorting Ethel ashore where Dr. Crippen is already being held. He'll be taken back to England, tried and hanged.

CANADIAN PACIFIC RAILWAY
STEAMSHIP LINES LIVERPOOL

AM VERY SUSPICIOUS OF TWO PASSENGERS WHO ARE OFFICIALLY MR. ROBINSON AND SON. THEY SEEM TO FIT INSPECTOR DREW'S DESCRIPTIONS OF DR. CRIPPEN AND ETHEL LE NEVE. OUR ETA MONTREAL IS 0900 AUGUST 2ND.

KENDALL

The CPR told Drew. Drew booked a passage to Canada (with two other police officers) in the *Laurentic* -- a faster ship than the *Montrose*. Even though the *Montrose* had a two-day head start, the other ship would get there first. Drew told the newspapers. For the next week everyone on two continents would have one topic of conversation. Charts and diagrams would show how the *Laurentic* was catching up then passing the *Montrose*. No one on the *Montrose*, except Captain Kendall and Sparks, knew what was happening. Kendall took a picture of the guilty couple striding down the foredeck (see photo).

As the ship steamed up to the Montreal pilot-station (Father Point, St. Lawrence Seaway) four "pilots" -- only one genuine -- in blue suits and white caps boarded her. Chief Inspector Drew arrested Dr. Crippen. He was returned to England, tried, hanged.

The British government gave a £250 reward to Captain Kendall for bringing Crippen to justice. That was not the end of Kendall's story. Nearly three years later (May 29, 1914) he was captain of the previously mentioned* *Empress of Ireland* when she collided with the Norwegian vessel *Storstad* at almost the exact spot where Crippen had been captured. There was fire and tremendous loss of life (1,023 died, 444 survived). Kendall went down with his ship but surfaced and clung to wreckage -- in the same way as Ron Ferguson, his Sparks. Half an hour later they were rescued by ships that had been brought to the scene -- by wireless. (The eight seconds Ferguson spent sending an SOS were productive!)

---

*Page 60.

*****

VERY EARLY RADIO SHACKS

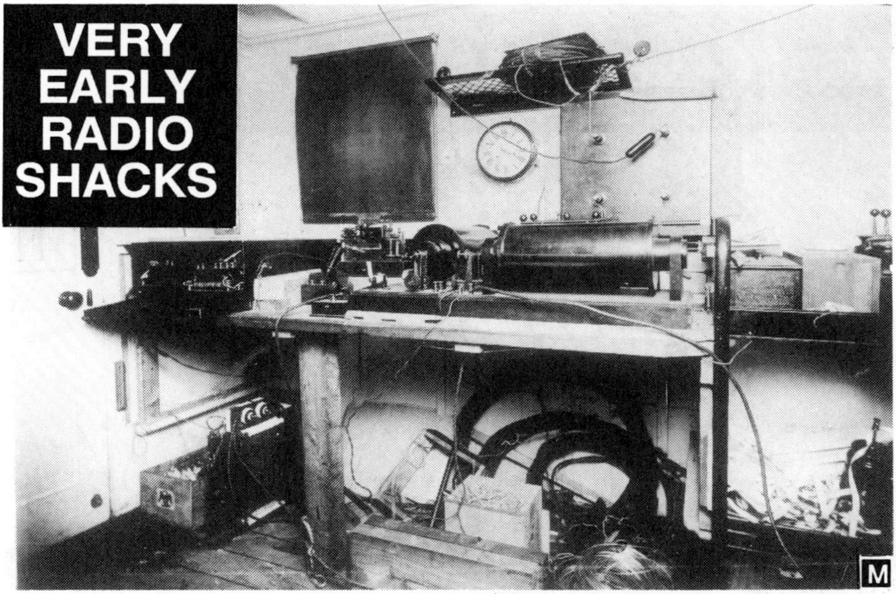

A very early radio-shack — maybe *the* original — in spare corner of ship. Condenser, induction-coil etc. King Clock is present!

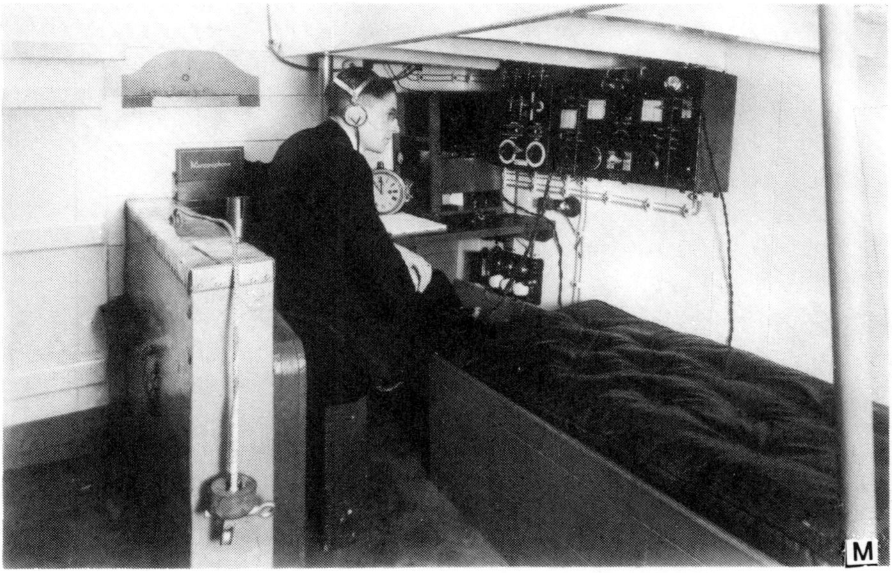

One of the original Sparks in his radio-shack. Coils, spark-gap, earphones, etc. All crude but working — plus alarm clock!

AGAIN – VERY EARLY DAYS

Transmitting equipment at Poldhu (pronounced POL-jew), 1901

Princess Clementine. **One of the first ships to have wireless. The cabin for this built between funnels.**

M

1903 — The very first radio school Frinton-on-Sea, Essex, England

EARLY - BUT NOT QUITE SO CRUDE - RADIO SHACKS

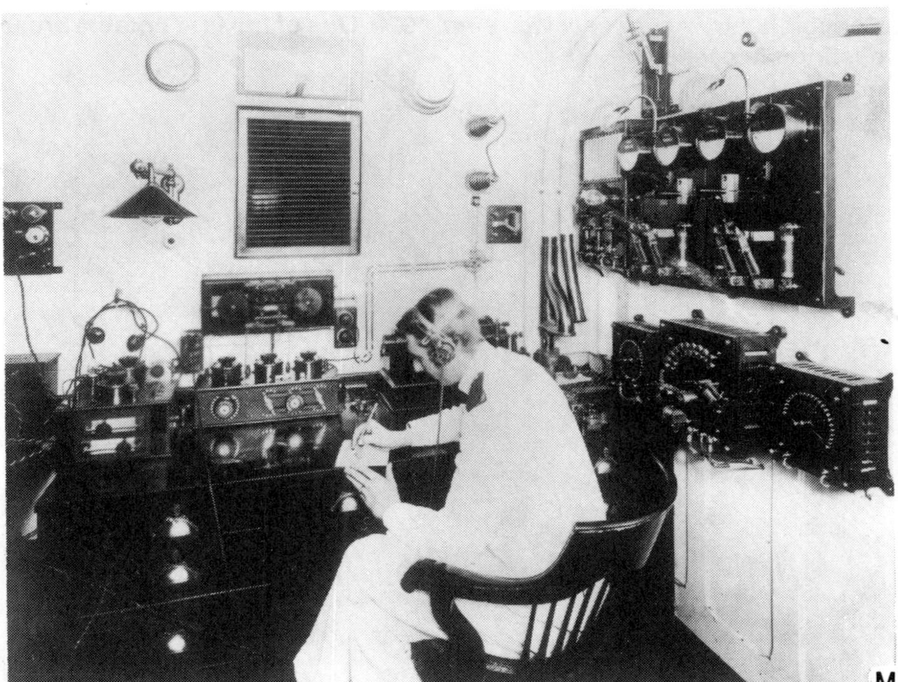

A typical Marconi Marine radio room installation in the year 1912. Almost certainly the appearance of the *Titanic* radio shack.

Concert heard mid-ocean, *Victorian*, 1920. One of the first notable broadcasting successes.

Radio Shack 1938

RADAR

The "scanner" high up on deck. Developed just before World War II. Great help locating German aircraft in Battle of Britain.

A NAVIGATIONAL-AID UNTHINKABLE TO OLD-TIME SAILORS

Radar reduces to a picture all solids (ships, rocks, buildings, etc.) within 40 mile radius. Navigating Officer sees everything (all solids) on screen.

DIRECTION-FINDER Where the two "lines of direction" cross — that's the "fix". That's where the ship is.

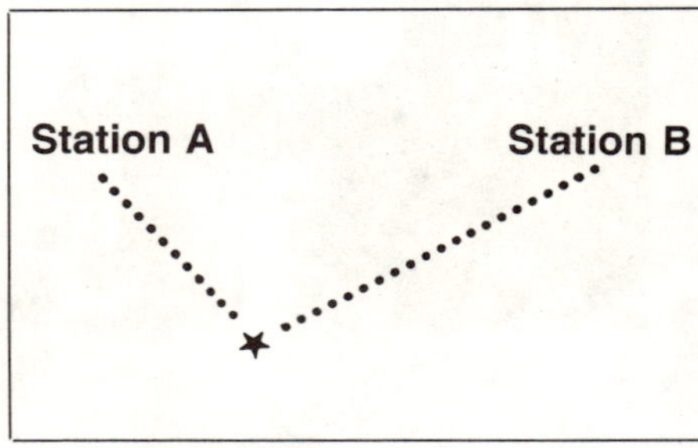

Navigating officer using Direction-finder in chart room.

◀ (King Clock has SPs for voice/ morse.)

At each hour/½-hour the R/O listens for "MAYDAY" ("M'aidez" --- help me!)

At ¼-to and ¼-past each hour, he listens for SOS (· · · - - - · · · easy, distinctive combination of dots & dashes.)

A radio console 1970-1980.

SOME SYMBOLS EXPLAINED

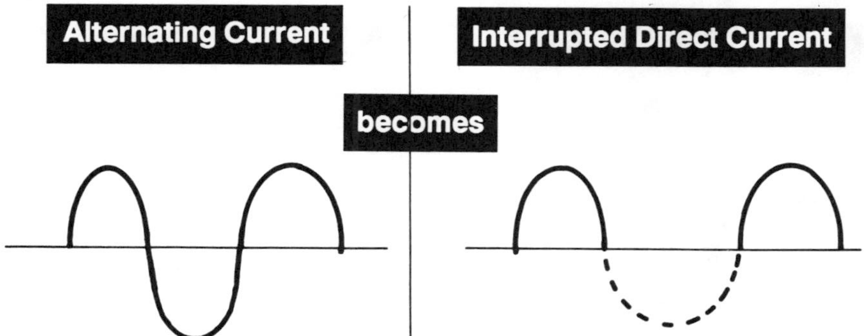

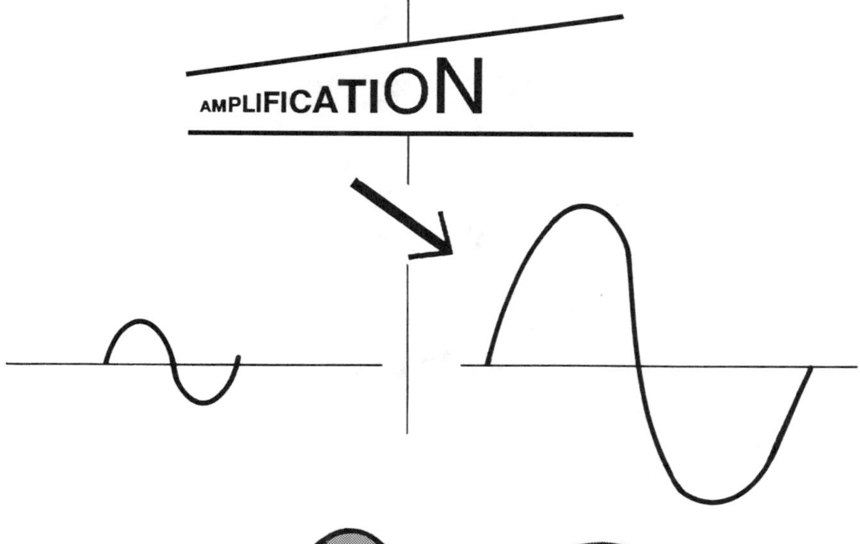

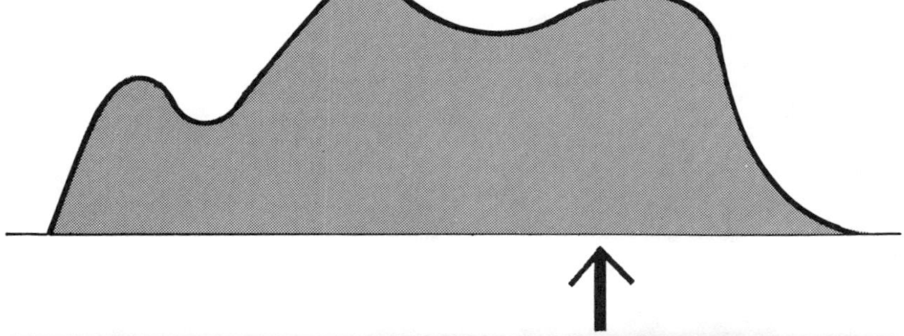

AC becomes DC, then is "modulated" by speech/music

"PRESS, O TROUSERS!"

In the summer and fall of the year 1940 when England stood alone against Hitler, the RAF fighter pilots (directed by Air Marshal Dowding) stopped the rot.

Little things can eclipse bigger ones by their importance -- or proximity, anyway. Trivialities get a sort of passport to immortality just through association with that time.... Cambridge was occupied by the RAF aircrew-to-be. We were billeted in all the various ancient colleges. A golden misty nostalgia steams up from anyone or anything connected with that place or period, even from the tyrannical little flight sergeant whose squads won the drill cup week after week. I was in one of them.

I was billeted in Pembroke College. My roommates were Ron Somebody and Somebody Jolliffe. Ron, miffed and jealous over some squabble about a girl, locked me out of our secret cellar-entrance. I had to spend the night in an air-raid shelter to avoid being nabbed by police at the college entrance. Can't remember Jolliffe's first name, but I see him vividly in my mind's eye. He always had a half-smile and a chuckling approach to everything.

It was the night before the drill competition. Our sergeant was expecting us to win for him again. We were 20 "erks" (AC2s--the lowest form of animal life in the RAF) but together we made a well-oiled machine. We now had to look perfect. Buttons and shoes had to be polished until their gleam made you blink. Haircuts, clean uniforms -- all had to be just so.

Ron, Jolliffe and I had no iron to press our trousers, so we put them on our mattresses, covered them with blankets and sheets, and hoped that the weight of our bodies (as we slept) would do the trick.

When we turned in and put the lights out, I was on one side of the student's bedroom the RAF had commandeered, Ron and Jolliffe on the other. Out of the dark, Jolliffe's voice sang out:

"Press, O trousers!"

(We won, next day, as usual.)

Trivial. Silly. Undeserving of such diamond-hard permanence. But there it is.

* * * * *

It's 1980. The bright, noisy here-and-now is Callao, Peru. The busy port is full of ships. It's evening, and all their lights are on. Down at the dockside bars, there is dance music, and the prostitutes are waiting, hopeful.

I get into my bunk, and turn off the light. I hear Jolliffe's earnest desire that our trousers should have a knife-edge to help us win the drill cup.

A SPARKS IS BORN

I seemed to be drowning in the cataract of Morse flooding through the earphones. It was over 30 words a minute -- a little faster than I could comfortably handle. As I struggled to keep up, I was going under, coming up, gasping for air: just like drowning. It was my stint on the *City of Canterbury* to get the press report at midnight for the ship's newspaper. The mess I made gave the chief radio officer some hard guesswork (almost illegible, frantic scrawls).

The ship was plowing over the equator, headed for South Africa from London. I was on my first ship as a radio operator. I'd been waiting 18 months with my license for the Marconi Company to call me. I was one of three operators on the ship, and mine was the graveyard shift. I just wasn't used to high-speed, plain-language Morse, and here I was, very unhappy, frantically darting about the typewriter keyboard.

My nervousness washed over onto everything else I said or did. I was labeled "odd." This made me a scapegoat, a punch bag, a source of ridicule to one of the young cadets. He was fresh from a maritime school, and couldn't abide anyone who didn't fit in with his idea of what was right and proper.

So, between the frantic mood induced by the Morse and writhing under the cadet's supercilious scorn, I was not just miserable; my mind was gyrating anticlockwise, in and down, round and around. I didn't know how I was going to stick it out, down to the Cape of Good Hope, to South African ports, up and down south-east Africa, back to England.

The nadir of this mental distress came one day when, alone in my cabin, misery seemed to be exploding upward from my guts. I had to cover my face with my hands to stop myself from flying into pieces. After a few earthquake-like sobs, I moved my hands away from my face and saw my Goanese steward leaving the cabin. He'd seen me, crying like a kid! I expected the news to travel quickly. The supercilious cadet would *really* have something to play with. This little incident occurred while the ship was en route from Capetown to Durban. It crystallized my growing determination to get off the ship as soon as possible, to get reassigned by the Marconi Company to another ship.

Why, you ask, was I weeping?

Capetown had been a very welcome change . . . table mountain over all . . . different . . . Dutch-clean . . . colorful sights, especially the Zulu warriors in their weird costumes, with huge ostrich feathers on their heads, pulling two-wheeled "taxis" and begging for business. I had gone on a pub-crawl, sampling the excellent South Africa wines. I made acquaintances, sitting on my barstool. I expatiated on life at sea and life in England and all the things that were as strange to these Afrikaaners as they were to me.

What I didn't know at the time was that a pretty little South African damsel had been soaking up my every word, longing to catch my attention, to speak to me. How could I know this now, if I was unaware at the time? Because when I got back to the ship later that night, they (including the contemptuous cadet) told me, in amazement, that some girl had been telephoning the ship, asking for me. (The ship, on arrival, had been hooked up to a shoreside phone.) She'd seen me in some shoreside bar, and had been unable to get my attention. They gave me her name and phone number to call. I wondered if this was some practical joke, but the mouth-gaping wonderment of the cadet told of his stupefaction and jealousy at this rare occurrence -- a woman chasing a man.

Rare? Well, new to me! And astounding to the cadet. Among animals of many kinds, and in the human species when it does occur, the female in heat is single-minded, ruthless. How much so? I got some idea when I called her and spoke to her for a while. I was amazed at the information she'd unearthed and made use of, and at the plan she'd devised with Machiavellian ruthlessness.

She'd found out that the ship was leaving at 10:00 AM the following day for Durban (about 750 miles up the coast) and we had no time to get together. However, she was going to *drive* to Durban, and she'd meet me at the Royal Hotel at 7:00 PM the night after next. She was sending a letter to me on the ship by hand mail. I'd get it the next morning before we left. It would have her picture and other details.

Phheew! I collapsed in my cabin and began to laugh. This girl (she'd sounded sweet and pleasant on the phone) had done plenty for me already, even if I never met her. She'd blown me out of the self-pitying mood in which I felt helpless, unable to act in self-defense. I spent the evening wondering about this gal who had a crush on me. Next morning her letter arrived, brought by some special carrier. She'd worked out everything -- she was *determined* and crafty! Inside was a snapshot.

She was a pretty little thing, slim, shapely, eyes full of fun, standing by a garden gate. I read her neat handwriting. She was 18. Didn't want me to write to her because her parents might see it -- or at least the envelope and strange writing. (What excuse was she going to give them for suddenly jumping into a car and making a big cross-country drive?)

Four Bells. Ten o'clock. "Visitors ashore!" Sailors rushing around. Gangway down. Ropes cast off. Bells again. All the bustle of departure seemed to say, "We're real. Not this fluffy little girl with one idea in her head." This mood was reinforced by the smothering hand of the ship's routine coming back so quickly, all over me like a giant's hand. The 30-wpm Morse news had to be taken at 11:00 that night, and by that time I was weak, despondent, an easy prey for the all-over feeling of defeat as I struggled with the crisp, remorselessly-advancing Morse. Miserable feelings piled up. My unpopularity, my general malaise, weeks of self-pity as we'd sailed to, over and past the equator. Now, I was stuck with a might-have-been, a nearly-but-not-quite that *might* (but probably wouldn't) lead to an interesting rendezvous in Durban. She wouldn't come. She'd get cold feet after her sudden infatuation. These were the negative thoughts that swelled into the mental agony I was in when my bedroom steward found me sobbing.

The ship got to Durban early, two days later. I went to see the manager of the local Marconi depot. I pleaded to be released from the *City of Canterbury*. He was sympathetic, said there *might* be a chance for me coming up next week. The R/O of the *Silver Oak* had been pestering him for months for a relief. He'd been going back and forth from South Africa to the West Coast of the USA for over a year now, and he wanted to go home to England. The Marconi man said to me, "The movements of the *Silver Oak* might coincide with your ship. Where do you go from here?"

"To Lourenco Marques," I told him (dare I hope?), "then Beira; then we go back port-hopping the way we came. Lourenco Marques, back to here, Port Elizabeth, Capetown -- and Portsmouth, England."

"I'll follow your ins and outs. I *might* be able to fix it -- don't hope too much."

I decided to hope. It kept me going for the rest of the day: desultory sightseeing, a bit of boozing, then back to the ship. Everything focused on 7:00 PM, when Betty (that was her name) was due to meet me at the Royal Hotel. I forget how it happened, but the vast horrible fact was ... I was LATE. Not very much -- 20 minutes or so. She wasn't there. She didn't come. I chain-smoked. I drank beer after beer. I looked around the hotel. I was sucked by an anticlockwise motion of a vortex down and down into misery. I never did meet her, nor hear from her again.

A tragicomic little incident, you say. Yes, you, O Reader! I'm inviting you to step into and through the page like Alice Through the Looking Glass. For lo! Now, these many years the arch-habit of addressing the Reader personally has been swept away with other silly 19th-century gimmicks, like flowery Gothic architecture and antimacassars. But there's something in the underlying idea. We meet, beyond these printed words, beyond Time and Space, look down objectively at the dejected chain-smoker in the Royal Hotel in August, 1950, and sympathize with his restricted viewpoint, boxed in by Unknowns....

Easy to be detached and philosophize from another point in the space-time continuum, but my viewpoint at that time got lodged, fixed for years. It was MY fault, for being late. Alternatives such as her standing me up (deliberately or accidentally) didn't even occur to me. Twenty minutes wasn't much for a date made 750 miles away, three days earlier, by such a very, VERY (at the time) determined young lady. She'd have waited.

Ah, that's common sense talking. Not me, the way I was, and the way I thought about it for years. She came. I wasn't there. She left. Just like that. It was my fault. Those alternatives, as I keep saying, didn't even suggest themselves, for years. About two years later it occurred to me that she might have chickened out. Ten years later, on my first return visit to Durban on another ship, when I nostalgically roamed and of course went to the Royal Hotel, it occurred to me that maybe *she* had been late. Maybe she came at 8:35, five minutes after I'd left.

To you it's just an amusing anecdote. You've already got the point. You want to rush on -- but to me, at that time, it was intense suffering!

Twenty-five years later it occurred to me that Betty had cooled off just a little bit, and had seen that she could get to me without so much sweat, dashing across the country, deceiving parents, etc., etc.--by simply sitting tight, and waiting for the ship to return to Capetown! That may have been exactly what she did. After being told by the local newspaper when the *City of Canterbury* would arrive back in Capetown, she would have been readyreadyready to telephone the ship, ask for me, and be told that I was no longer aboard. (I'd left the ship in Durban.)

Poor Betty. I hope she has been a contented wife and happy mother for a long time, and has perhaps only recently become a grandmother. What fun if we could meet or at least correspond on the subject of What Happened and Why at the Royal Hotel, August 12, 1950.

Speaking of those events, there I was, looking down a narrow tunnel in space/time, seeing only that the ship's going on to

Lourenco Marques has made a mess of a Possible Date with a Gorgeous Woman, and there's only a faint hope of another ship.

In my misery and desire to escape, I got as near to it as I could by going ashore in Lourenco Marques and staying overnight in a hotel. Vivid still is the memory of a slowly turning electric fan seen through the mosquito netting surrounding my bed. When I returned to the ship, Chief R/O Lovelace told me that they'd informed the police to look out for me. Where the hell had I been? I hadn't told anyone I was staying ashore, etc. I shrugged my shoulders. I couldn't care less. I was so low, one section of the barrel-bottom wasn't different from another part a few inches away.

O Reader, what sort of claptrap or miserable self-pitying little tale have you landed in? I've exposed myself as an incompetent crybaby, a would-be Sparks who can't even be relied on for some high-speed Morse, can't get along with his fellows, and not only makes a backhanded idiot of himself over a silly little girl in a port, but interminably bores you by going on and on and on about her. What are the Marconi VIPs and shipping companies going to do with this fellow? Best thing is to get him to London, pay him, run him out of the British merchant marine forever. Is that what you're thinking? Well, so was I.

Relief, if not happiness, was round the corner. On return to Durban, the Marconi manager said the *Silver Oak* was in port, waiting for ME to come aboard, get introduced to the operator, the gear, the captain -- prior to swapping positions with the wireless operator of said ship!

I was introduced to Johnny Somebody-or-Other, the wireless operator of the *Silver Oak*. A nice guy. The manager said, "Let's go," and go we did. No one could help being impressed by her appearance. Modern, streamlined, colorful, she seemed built for speed and efficiency. Aboard, the first impression was reinforced: woodwork, paint, gleaming brass. The long, sumptuous staircase gave the ship the atmosphere of a first-class hotel. We ascended a narrow staircase, past the captain's cabin. ("We'll see him later," said the manager. "Come on up to the radio shack.") Johnny led the way, opened the door.

To say I was impressed would be a gross understatement. I was divided between being wonder-struck and shit-scared! Knobs! Switches! Buttons! Meters! I'd never seen so many controls. (Later I counted them for the hell of it: 121!) Transmitters, receivers, a D/F* wheel, the auto-alarm, boxes of gray steel with little knobs, big knobs, controls that I was supposed to know all about, gimmicky things galore behind which I would be utterly alone, across the vast Pacific to America, half the world away. I gulped. As Johnny explained this and that, his voice came through a mist from a long way away. This was crazy. I'd better call the whole thing off. (I didn't say this, but my thoughts shouted it!)

I don't belong here . . . I haven't much more than a clue. Gosh, I've never even hit a live Morse key with my hand in my life! Lovelace didn't trust me, just kept me on 500 khz all day, then that vile press report at night.

Somehow, I gathered enough self-command to listen to

---

*Direction-Finder

what the manager was trying to tell me. "Come on. Start the transmitter. Here's the motor handle."

He was pointing under the table in front of the main receiver, between it and the operator's chair. There was a black knob attached to a sliding mechanism that probably made a motor generator work. I wasn't sure. I'd never seen one quite like it before. I grabbed the knob, pushed forward hard.

The motor generator screamed like an angry sea gull, and was cut off by a big bang, while some fuse literally hissed as it expired. Johnny and the manager were rolling with laughter. The starter was supposed to be pushed forward gently, up to the first of a series of metal studs. When it had gathered enough speed there, I should have then pushed on to the next stud, and let the speed get steady (like a series of gears). I had pushed the starter through six studs; a blown fuse had prevented any worse damage. "You'll learn," chuckled the Marconi manager, "the hard way."

The captain, I heard, was a peppery old curmudgeon called Morgan. The Marconi manager said he'd tackle him about me. The Old Man had exploded "in a mixture of fright and fury" (I was told) when he glanced at my seaman's papers and saw that I'd had under six weeks at sea, with no experience as a lone operator at all. The manager said that Morgan had burbled on about all the ports from India to Singapore and on to Manila even before his ship began the long, long Pacific passage to San Francisco, the bad weather he expected, the risk of so many lives and the safety of the ship, etc.

"I had quite a job calming him down," said Mr. Thompson, the Marconi manager to whom I was beholden, and grateful, for the opportunity. "I said you could be trusted. Don't let me down." I was going to do my best, but I hoped I wouldn't have to learn *everything* the hard way. This Captain Morgan seemed to be a prickly handful to cope with, and there'd be nobody to help me, 3,000 miles from anywhere, in mid-Pacific. I still hadn't met him.

Johnny patiently (through my cloud of nervous uncertainties) explained how all the gear worked, how to do this and that, where everything was, what stations to contact for so-and-so, what to beware of in this place and that, etc., etc. After we'd arranged to

The first radio shack where I was in complete control. You know little about radio? Me Too! (then)

If you "know little about radio," it's easy to get your sympathy. You look at the picture with the bewilderment I had at the time. Then, if you jump into the picture, see it as real, and suppose that you're really going to be put in charge here–you'll experience my extreme uneasiness, if not terror! Even if you know a lot about radio, you'll appreciate my feelings as a tyro! Well, it was a long time ago, but the scene is acid-etched into my memory.

① Please, to be truly "simpatico," imagine **you** are an ignorant, fledgling Sparks with no experience whatever.

② You can send/receive Morse . . . but THAT'S ALL.

③ You've just been introduced to this scene. You're going to be here, **alone**, across the Indian Ocean and clear across the Pacific.

④ How do you feel? What's going on in your mind?

⑤ Coarse vernacular best expresses your feelings: You "shit blue lights!"

⑥ You were the original artist. (Oh, let's stop pretending ... it was I, the author.) It was tidied up by professional artists.

a. A few of the many books/instruction manuals/official regulations, etc.
b. The Medium Frequency Transmitter
c. The High Frequency Transmitter
d. Antenna-switching board
e. Emergency transmitter
f. Typewriter
g. Various gimmicks & machines; their uses forgotten!
h. Main receiver
i. "The Stabilizer" – teapot with lifeline to sanity!
j. Direction-Finder (D/F)
k. Receiver connected to D/F
l. The Boss! The GMT clock (with Silent Periods marked in red) dictating all schedules
m. Battery-charging boards
n. Auto-Alarm
o. Voltmeter
p. Bell-push to bridge
q. Morse key
r. Main-Motor starter (Works by a series of studs, gradually increasing speed), SEE STORY ABOUT THIS
s. Power regulator
t. Fuse box
u. Emergency motor starter
v. Power regulator? (not sure)
w. ??/(forgotten)
x. Ear-phones
y. Voice-tube to Bridge

rendezvous on 16 mhz at 1530 GMT (when British ships were in long-distance communication), he wished me luck, and went.

Luck. How big a slice of it was I going to need? I looked at all the knobs and switches in this tidy, efficient-looking cabin. This was to be the center of everything for me, through 8,000 miles to the USA and back. Captain Morgan would turn to me when he was thirsting for information about cargo plans, shipping agents, storms, what the head office wanted, navigational hazards and so on and so forth. I'd better familiarize myself with everything in the radio shack. I could at least get some practice at operating the main transmitter. So (carefully and slowly this time!) I started the motor generator.

It *was* a noisy thing (this was before the days of the silent, transformer-started power supplies) and when the handle was on the last stud, the desk vibrated to the loud hum and rattle. Then I thought I heard someone shouting outside the radio shack, but I couldn't be sure. I opened the door and looked down the short flight of steps to the deck below, to see a large, fat man, with a furious expression on his florid complexion, glaring up at me, yelling, "Hey -- Sparks!" (Ye gods, it must be the captain!) "TURN THAT DAMNED THING OFF!"

"Sorry, sir." (He disappears. I turn around, pull back the starting handle, flop into a chair.) This little incident was a sort of prism through which I could see a grave error in the design of the ship.

That noisy motor generator was right over the Old Man's cabin! Every time I used it, his cabin would shake and reverberate! Even an experienced, acceptable radio officer would to him be just another of those vile monstrosities of the Sparks species, who could and would disturb his peace and quiet many, many times daily. Not only that, but this tyro, this damned beginner that the Marconi man had foisted on him, hadn't even learned to respect a period that is something akin to holy aboard ship -- the siesta hour!

Well, we both survived. While I tried, but made little progress in getting along with him, he (ignorant and unappreciative of that) stagnated in his prejudice against me. As we sailed up the Indian Ocean, north-northeastward to Bombay, I began to get the hang of the electronic contraptions all around me, to become comfortably familiar on the air with other ships and coast stations. My window on the world was not that of sight but of sound -- Morse cracklings that gave a mental picture of fellow operators in South Africa, Portuguese East Africa, India and on other ships. The headphones and the Morse key were just crude gateways to another world, not restricted to the ship's boiler-fired propeller shafts to drive the ship at 15 knots, but zipping messages back and forth at the speed of light, 186,000 miles a second.

On the HF Sked (high-frequency schedule) I contacted Johnny on the *City of Canterbury*. They were on the same latitude as us roughly, but in the South Atlantic. We talked to each other across Africa. In the Morse jargon with which I was becoming familiar, he said HW OM? ("How are you, old man?") HW U DOING? To which I replied, OK OK, but we lost contact through heavy atmospherics and signal-fading.

I *was* doing okay, too, with regard to self-management, familiarity with all the knobs, the strict schedules of so many stations, the peculiarities of the radio receiver when tuning in, where to find this, that and the other, keeping the 500 khz watch as the basis for everything, and contacting other ships, all nationalities, either to pass the time of day by swapping positions and movements or giving each other vital info. I was making great progress and getting comfortable.

Yet Captain Morgan, in his prejudice against me, was the precursor of several more Old Men I was to know. That frame of mind is blind to evidence: the good things are not seen; prejudices are enlarged, multiplied. I was no good; I couldn't be any good; I was a thorn in his side. These general ideas of me he had spread to the other officers. It was a self-feeding vicious circle. He detested me and showed it. This made me nervous and overreactive. This was seen, heard, felt by others; they developed a dislike for me that was both self-feeding and helped by my reactions to them.

I had a friend, one of the junior engineers. He would come up to the radio shack when he came off watch to chat. Or when in port we might go around together. When he came into the officers' mess, he'd wave a cheery "Hello" from the opposite side of the room, where his place at table was.

The seating arrangement was this: all the deck officers were on one side (I was at the bottom of one table) and the engineers (with my friend) were on the other. At the head of my table was the ship's purser, a man who has a position that can be peculiar and indeterminate, according to the ship/the time/the company and other factors. On a passenger ship he is a VIP, the go-between for everybody to/from everybody else. On some cargo ships he might be regarded as a sort of general manager, the one who looks after the stores, mainly. For this job, he might be elevated to a position alongside the officers, or he might be with the crew, and be both treated and behave like a general factotum. On other, in fact most, ships, he doesn't even exist! Those jobs of accounts, bookkeeping, stores management can be split up between the captain, the chief mate, the radio officer and/or any third mate. Or even an apprentice (use him as slave labor, with no recompense). So a purser can be a VIP or a menial, a manager or a superfluous nobody. The purser of the *Silver Oak* came in the first category. He was the Apple of the Old Man's Eye. Before I say any more about him, though, we must finish going round the eating places.

The captain didn't eat with the other officers. He ate in another room, next to the officers' mess, with only the chief engineer for company. It was most peculiar. I've never seen anything like it, nor even heard of it, any ship, any time.

However, when ashore, Captain Morgan was buddy-buddy with the purser. Once when I went ashore in Singapore, I was eating in a restaurant when I looked up and saw the Old Man and the purser in a far corner, wining and dining with great conviviality.

Well, I vacillated between being just as petty and small-minded as some other people aboard and the feeling I began increasingly to have in the radio shack. Why, here I was king, godammit--dictator, government and people. There was no competitor. The Old Man might hate my guts, but he had to come to me to send a message or take what I gave him as messages from others. I was Somebody.

The little radio stations in the Pacific gave me problems. Sometimes their English was very shaky; sometimes they weren't available when and where they should have been; sometimes the

Old Man had to send or receive some vital info, but several other ships would be ahead of me in the lineup to speak to a radio station. Somehow, everything got ironed out. One night, just before going off watch in the evening, I strolled on deck to appreciate the vast, star-studded sky of the South Pacific, where the Milky Way is so much brighter than it is in the North, and I felt happy, adjusted, at peace. In this vast universe, where we could only see our galaxy, 100,000 light years across, this little speck, the Earth, had its place. This little ship was coming into the Pacific Ocean (three times bigger than the Atlantic) and we were heading for San Francisco. I had control over everything in that radio department, which was nothing less than the Ears and the Voice of the ship. Yeah, no wonder the Old Man had got scared (that was it--he was just scared), scared of being alone and incommunicado on this vast ocean if/when his inexperienced communications officer let him down. I *almost* felt like being sympathetic to the Old Man, who after all was the master of the ship and therefore responsible for every damn thing and person and event on it. In a democratic country, a ship's captain is the only official dictator you can have....

What the hell was I thinking? I didn't want to be a devil's advocate -- I wanted to get into the radio shack, hit the key, and talk to people! There were people all over: ZDK (New Zealand), VPI (Hong Kong), even KFS (San Francisco), whom I spoke to last night, though we wouldn't be there by ship for another 18 days. I was Morse-key happy, Heaviside-Layer happy (this layer is 150 miles above the earth; shortwave transmissions bounce off it). I shared the happy freedom of radio amateurs everywhere and fellow Sparkses everywhere. The world was our oyster!

I was getting quite familiar with the equator. I'd crossed it now four times, going north or south, though I'd only been at sea about three months. (The first time, on a passenger ship, they'd made a big deal of it, of course--Father Neptune, dunking people in the pool, etc.) My introduction to the dateline came when the second mate pinned a notice on the board. "Today was Tuesday, 17th September. Here's your chance to relive the past. Tomorrow is Tuesday, 17th September, also! We cross dateline tonight. Clocks back 24 hours."

Shuffling around with time began to have another meaning for me when I started taking an interest in the stars. The second mate introduced me to some constellations on the middle watch (midnight to 4:00 AM, when nothing much is happening and most people are asleep). We only got semi-friendly. The second mate was too close to the Old Man, and was therefore affected by the poor opinion the O.M. had of my character, ability, trustworthiness and everything else. But we talked. He got me familiar with Orion's Belt. I was interested in the light-years distance away. The mate had to consult some books in the chart room for that--that sort of info wasn't required for navigation. I learned that the top star of Orion's Belt, Betelgeuse, was 680 light-years away. That meant we were looking at light that had left it in the year 1270 A.D. The middle stars were 2,000 light-years away. The bottom one, silver Rigel, was 900 light-years away. Its light had been traveling to us since the year 1050 A.D.

We were getting near the end of the eastward journey across the Pacific. I'd been busy on the direction-finder in foggy weather. I got a break and was reading *Worlds in Collision*. There seemed to be scientific evidence to support the theory that the world had once spun the other way, and the sun had risen in the west. In 647 B.C. the planet Venus had skirted near the Earth, and had made it revolve the other way. Chinese, Peruvian Indian and other texts, including the Old Testament, were quoted in support of this theory (Joshua making the sun stand still was the Old Testament bit). I fell asleep reading it.

I was awakened by a very weird, off-key foghorn. The ship seemed very still and quiet -- no engines -- we had stopped somewhere. I peered through the porthole. Fog swirled up and around. The weird, ghastly foghorn wailed every minute. Of course! We'd arrived! This was San Francisco Bay! That foghorn was Alcatraz! (At that time, it was still used to imprison desperate criminals for long terms, mostly life sentences.) That was my introduction to the USA.

My only memory of San Francisco at that time is of a pretty dental nurse bending over me as I got some dental work done, halfway up a skyscraper. This may have been some sort of premonition about a Dutch dental nurse I was to fall in love with four years later, but it doesn't say much for the glamor of that fascinating city. I became much more familiar with it later on.

Tacoma, Seattle, Vancouver -- I was in a sort of fairyland, where everything was so new to me and interesting. All along the West Coast I was captivated by the style of living, the way folk talked, the *differentness* of everything and everybody.

Back on board, heading for the Philippines, I was immediately back on the pendulum, swinging from the fun and interesting life in the radio shack (contacting people and ships many, many miles away) to the disapproval or grudging forbearance -- some version of my unpopularity -- with the captain and his navigating officers (denizens of the bridge, to which the access was a sort of cubbyhole from the radio shack, to my far right).

I decided I had to make a good impression somehow. The night before we sailed I had sneaked into the wheelhouse to take a little fuse out of the depth-sounder; they would have to call Sparks to help (it was an electronic machine that came under my jurisdiction) and I would be The Hero, The Rescuer. Well, they called me when the machine wouldn't work, and I spent some time "investigating," testing this and that. (There had to be tension, uncertainty, delay, etc.) I slipped the fuse back in position and let the captain and mates know I'd "fixed" it. It didn't raise my status or improve my standing in their eyes one little bit.

Then, as if I hadn't already been unspeakably low in their estimation already, one incident made sure that I was one of the equivalents of India's "untouchables." We were in Vancouver. It was shortly after Christmas. (The captain had had a Christmas tree in his room, and I with all the officers had had drinks there with Yuletide spirit. He had even shaken hands with me!) There were some visitors aboard. (Forget who they were. Doesn't matter.) There we were, in the officers' mess; and I, in some conversation, said very loudly that we were "all fucked up!" I suddenly remembered the visitors, women among them, but it was too late.

Get the parameters right. This was 1950, before the permissive society for sex/language, etc., really got going. It wasn't anywhere in the USA -- where a certain tradition of "anything goes" was part of the ambience. This was a British ship on British territory, and people were strait-laced and stuffy by tradition. My language caused a shocked silence even greater than that which had greeted Eliza Doolittle's "Not bloody likely!" among the VIPs in the paddock at Ascot.

So my social standing remained irreparably low, but my confidence on the air grew. Over the Morse key, managing the whole radio station of GCQR (the ship's call sign) and even making repairs (genuine this time), I recrossed the Pacific, traveled through the Philippines, called on Burma and came into Calcutta. My confidence was such that I could truly feel, after six months on the *Silver Oak*, that I was no longer a nervous beginner, but a true Sparks.

Captain Morgan's prejudice wasn't going to be dislodged, nor moved an inch, once his mind was made up. He could and did ignore (perhaps he just didn't see) whatever RIGHT things I did. All the telegrams, coming or going, were taken for granted. All the twice-daily weather reports (sometimes vital info on Pacific typhoons), all the routine, mostly boring things that made the radio station of the *Silver Oak* so praiseworthy and efficient just because there was nothing remarkable (i.e., newsworthy or disastrous) about it, all this he ignored. It was simply no evidence of my true worth as his communications officer. He was waiting for me to do something wrong, and also waiting till we got to a British port, where there would be a Marconi depot and he could get rid of me.

Captain Morgan, where are you now? (I met someone in a bar in Cochin, India about 10 years after my time on the *Silver Oak* and I was told that Morgan had retired, gone to New Zealand. If so, you're in your 80s.) Or are you dead? If so, that makes it easier to get in contact with you. It's your spirit I want to talk to. Now that we are detached from the exigencies of *Silver Oak*, I truly sympathize with you! I see your point of view completely. You had a 15,000-ton vessel to take safely across the Pacific, and you had to be *sure* of what to expect at every port. It wasn't fair to you to make you trust a raw beginner who had only been a few weeks at sea and had never had any direct personal control over any radio gear or any of the electronic gear on a ship. I'll sum you up, Captain Morgan, in a four-letter word (not quite the same as my celebrated gaffe before passengers in Vancouver) -- you were SHIT-SCARED! And I don't blame you! But you were also patient. You knew I had to make a real barn-burner eventually. I did.

I had a master's message for the owners in London. The British had a system of linking Commonwealth stations together for relaying messages. I could and should have relayed this important message via Sydney, Australia, a couple of thousand miles to the south, no distance at all for my shortwave (HF) transmitter. Instead, I chose to send it through Singapore (at that time Singapore was not yet independent). They took the message addressed to London and sent it via a circuitous land-line instead of by direct radio link (or, better, teleprinter). So the Old Man's message was delayed. Yours truly, always on the Old Man's shit-list, had at last given him something solid to work on. Captain Morgan complained to the Marconi manager in Calcutta.

This character was thin and bald. He compressed his lips together tightly. He always did the correct thing at the right time. He wouldn't have put me on the *Silver Oak* in a hundred years. He made that clear. "Thompson," (the Marconi manager at Durban) he said, "did wrong in placing you aboard."

I suppose he did, but thank you, Mr. Thompson. To the end of my days I'll hear you laughingly say (as you reached for the fuse I burned up), "You'll learn -- the hard way!"

What had happened to me -- thanks to Mr. Thompson -- would have been impossible elsewhere. In the U.S. Merchant Marine, for instance, the inviolable rule is that a beginner must have at least six months' experience under others before he's allowed to take charge of a ship by himself. Here I was, being jerked traumatically from nowhere into a job that required professional perfection. This was very unusual in the British merchant fleet also, even though their system is a little more elastic. The tradition there is that exceptional circumstances call for a little rule-breaking now and then. That's what happened with yours truly. Mr. Thompson took a big risk. Peering through the fog of my jitteriness, he reckoned I'd muddle through ... and I did.

My fond feelings for the *Silver Oak* have led me to relate my connection with her right up to the time when I was transferred to another Marconi ship in Calcutta. However, the period covering the "birth" of Sparks doesn't coincide neatly with the dates on which I joined/left the ship.

You could say the "birth pangs" began when I nervously lunged at the motor generator and busted a fuse. When could we say the "labor" was over? Well, my realization or consciousness of *fait accompli* might be narrowed down to the period when we first approached the USA, coming up to San Francisco in thick fog, around 2:00 AM. I was tired out from hours (off and on) of direction-finding for the navigating officers. I picked up *Worlds in Collision*, then dropped it, and mused on my own little wonder-world.... We had just crossed the Pacific Ocean. Perhaps it wasn't too boastful to say that I had supervised and operated our radio communication links with panache. I'd won through. I'd gotten the hang of a vital job. I could converse with ease (or difficulty, if conditions were bad) with American, Australian, British and Canadian coast-station operators, and with any ship within range. With the right feel and know-how, I was cozy up there in the ether. I had the knack of all those dismaying knobs and switches. There were official do's and don't's laid down by international bigwigs in Switzerland for all ships, and by lesser authorities who had command over me, such as the Marconi Company. When regulations/ paperwork were driving me scatty, I could get sympathy and help from a fellow Sparks, of any nationality. We'd converse in a peculiar language--GE OM QRK4 PSE U GOT KFS FREQS? (Good evening, Old Man. Your signal strength very good. Have you got San Francisco frequencies?) I fitted in. I belonged.

A few hours later, I awoke. It was after 4:00 AM. The Alcatraz foghorn was wailing. Our foghorn was tooting every minute also. We had arrived. Personal triumph -- unknown to, and unappreciated by, the fractious Old Man and his mates, unappreciated even by the crackerjack American and Canadian operators on the West Coast, because they took efficiency for granted. I -- Sparks -- had been born.

THE FLOATING CIRCUS

I'm lying on a deck chair, looking up at the stars. Each one lies a certain distance in light-years from the earth. With light traveling 5.8 billion miles per year, I can label each star by the year light started from it. Looking out into space I gaze through corridors of time.

Planets go around those stars, and life on some of those planets is probable. Superior civilizations are possible -- and they are as far from us as we are from them. So people near the 1087 AD star, focusing now on England, would see Norman soldiers chasing Saxon maidens in this newly conquered island. Those near the 1767 AD star, focusing now on Philadelphia, would see -- not as a movie, but happening NOW -- Ben Franklin puttering around his printing shop.

There's a 1967 star. From there, on July 16th, with a super-powerful telescope, the Santa Clara can be seen in the Pacific, heading for Colombia, having just passed through the Panama Canal. Narrowing down to the radio shack, the telescope would hover on a small human figure -- me. On this particular day, I was nearing the end of my morning watch, anticipating lunch....

I glanced up at the big clock over the radio console. It had two hour hands. One pointed to 11:45 local time. The other indicated 1645 GMT -- Greenwich Mean Time, the international reference point for radio, navigation, space flights, etc. We were ready to begin the Silence Period, or SP. By "we" I mean myself and every other ship's radio operator on the face of the earth. Nobody --and that means nobody -- is allowed to send Morse signals on the international calling frequency of 500 kilohertz from 15 to 18 minutes and 45 to 48 minutes past the hour, every hour, every day, unless he's in real trouble, needing immediate help.

The minute hand edged into the red radial triangle painted on the clock face. It was now 46 minutes past the hour. Usually the Silence Period was just that: silence, period! It was emphasized almost to the point of being a Holy Period. Then, right in the middle of it--at 11:46 and 17 seconds -- the main receiver started barking out a series of long dashes: DAAAAAAAAH... DAAAAAAAAH ... DAAAAAAAAH.

Heck, I thought, someone is sending the long (4 seconds) dashes which should precede an SOS. These signals activate auto alarms on all ships within range, about a thousand-mile radius. These auto alarms set godawful bells ringing in various places (the radio shack, the radio officer's cabin, and on the bridge) to alert the Sparks if he's listening or to tell people to call him if he isn't. I happened to be on watch and was able to get the direct call, so my bells didn't ring.

The DAAAAAAAAHs stopped. Several minutes passed. The Sparks of the vessel in distress was waiting to let everyone get on watch. Then the message came. It was loud and obviously nearby.

> SOS SOS SOS (all distress traffic starts thus) CQ CQ CQ (everyone) DE (from) ZDKM ZDKM ZDKM (this call sign showed that the ship was registered in Gibraltar) SS COLON (ship's name) POSN (position) APPROX 6.20 NORTH 78.24 WEST STOP WE ARE WITHOUT POWER AND DRIFTING TOWARD ROCKS STOP 53 PEOPLE ABOARD STOP HAD FIRE IN ENGINE ROOM STOP NOW DISABLED STOP VESSELS IN VICINITY PLEASE INDICATE AND ASSIST IF POSSIBLE
>
> MASTER SS COLON

I blew into the voice-tube connection with the bridge. The third mate answered.

"Hello. Bridge." (A formal reply. He was a good lad, proud of his new authority, anxious to do the right thing.)

"Hi, Third. SOS message here. Please send someone to pick it up."

I switched on the power to the medium-frequency transmitter. Filament voltage registered OK. Plate voltage also. The frequency was set to 500 khz. After another 10-second warmup period, I was just going to turn the power switch to "max" when a fat, cheerful AB came to the door.

"Here, John. You must have flown down those steps! Thanks."

I gave him the distress message and hit the Morse key:

> SOS SOS SOS ZDKM ZDKM ZDKM DE WLGP WLGP WLGP (our ship's code letters) QSL (acknowledge) YR (your) SOS PSE (please) AS (stand by).

I then tried to contact them by voice on the VHF (very high frequency radio set -- good for about 40 miles if conditions OK).

"*Colon! Colon!* ZDKM Zulu Dog Kilo Mike. This is *Santa Clara* WLGP Whisky Love Golf Papa. Do you read?"

The static was deafening. I tried for a while. No good. I blew into the voice-tube to the bridge.

"Sorry, Third. Forgot to ask for our position."

"The Old Man has it, Sparks. He's coming down to you right now."

The main receiver barked "WLGP WLGP..." This was probably the *Colon* replying. I dropped the phone and sat down at the console again.

> SOS WLGP DE ZDKM MASTER SS SANTA CLARA... PLEASE GIVE US ASSISTANCE...CAN YOU TOW US TO BUENAVENTURA?
>
> MASTER SS COLON

As I was receiving this, the Old Man, Captain Lawson, had come in and was looking over my shoulder. As skipper, he was one of the best. I liked his aura of good-humored but firm authority.

"Tell him yes, Sparks. Here's our position."

He put a slip of paper on the desk.

"Say we can be with them in four hours. No, wait; I'd better write it down." He glanced at the clock, and scribbled, just to make it official for the books. I sent the message:

> SOS ZDKM DE WLGP MASTER COLON ... YES WE WILL TOW YOU TO BUENAVENTURA ... OUR POSITION IS 07.16N 79.21W ... OUR ETA YOUR POSITION ABT 2100GMT
>
> MASTER SANTA CLARA

"I can't get them on the VHF, Cap'n."

"I know. We've been trying on the bridge. I've got the chief firing us up on all the boilers. We're making 20 knots. Can't get there any faster."

Morse crackled out of the receiver again:

SOS WLGP DE ZDKM MASTER SS SANTA CLARA ... OUR POSITION VERY APPROXIMATE IT IS DR* ONLY WE HAVE BEEN DRIFTING SEVERAL HOURS. I HAVE RADAR EFFECTIVE RANGE 25 MILES. I WILL GET MY LINES READY AND TOW TO YOU WITH LIFEBOAT IS THIS OK WITH YOU?

ADAM WEINER MASTER COLON

"He's counting unhatched chickens, Sparks. We've got to locate him first. The Humboldt Current can take him anywhere if he's drifting."

"I'll get their Sparks to send a continuous signal on 480 khz. We can tune that in on the DF (direction-finder on the bridge)."

"Good idea."

It worked. The *Colon*'s signal was picked up at 124 degrees relative to the *Santa Clara*. If a cross-bearing could have been taken by someone else on another ship, from another angle, the exact position of the *Colon* could have been fixed, but this was good enough for now.

"Try to get New York on the single-sideband, Sparks. I've got to talk to my boss. The office is closed already. Here's Captain Bryant's home number."

No luck. Lots of static. When I did hear WOM/Miami, it was too unsteady to read.

"Sorry, Cap'n. The atmospheric conditions are bad. We're too near the Equator. Daytime heat. Might be better this evening."

"Hmmm." Captain Lawson took a reflective puff on his pipe.

---

*"DR" = "Dead Reckoning." "Dead" is "D'ed," abbreviation of "deduced." The known factors are speed, last known location, wind and current. From them, position is deduced. "DR" means "educated guess."

"Well, we'll send essential stuff by CW, but keep trying to get through by voice." (CW = continuous wave; the medium by which Morse telegraphists--including such eminent ones as Samuel Morse and Thomas Edison -- have been transmitting messages for more than a century.)

He knew the Grace Line wouldn't object to his obeying the First Law of the Sea -- to help others in trouble -- but he needed clarification about salvage. Our ship cost $10,000 per day to run (fuel, wages, upkeep, etc.), so time (i.e., money) spent in a rescue operation concerned us all. Whatever the salvage award was to be, it would be divided in the time-honored way; half to the company, half of the remainder to the captain, and the rest to officers and crew according to their rank.

However, our immediate concern was to reach the *Colon*. Communication was OK on 480 khz, by CW Morse. Talk (via the VHF microphone) was too weak and garbled; the sun scrambled it with static. Every now and then the Sparks of the *Colon* would hold his Morse key down for a minute or two, and the second mate of the *Santa Clara* would take a bearing on this signal with our direction-finder. Course and position would be noted, corrections made, and the rendezvous hastened.

All eyes on the bridge were peering through binoculars. John (the AB who'd collected the original SOS from me) saw the target first, and yelled, "There she is!"

"She's listing like hell," remarked the second mate. (Strong language from him, a quiet, reflective type.)

Silence while all adjusted their binoculars. Someone said: "What the heck do they have on deck?"

Everyone peered. Then, hesitantly, from the third mate: "Loo-looks like an *elephant*!"

Laughter, but then from the Old Man, quietly but decisively: "You're right."

Closer. Closer. Then the whole amazing sight was clear: Not one elephant, but two...and half a dozen zebras...and horses. Then the climax...two tigers in a cage!

Later information clarified things.

The *Colon* (only 2,226 tons gross weight, 305 feet long) had been carrying a circus to various South American ports. She was a "tramp" (snatching business if/when possible) but the British flag gave her a veneer of respectability.

As the *Santa Clara* approached, all the crew who were free to look (which meant everyone except those who had to work down below in the engine room, plus the six delighted passengers, all lined up on deck, taking pictures if they had cameras. The *Colon* looked very odd, listing over 12 degrees to starboard, topped by all those animals huddled on the foredeck.

This sideways lurching of the *Colon* certainly emphasized her crippled condition. Communications with her were also going to be crippled, as far as walkie-talkies went. Our sets were OK, but theirs were skimpy things, almost toys. Morse-talk was OK. We left our receivers on, tuned to 480 khz to get instant response. Even that was going to be limited, though: the *Colon* had no power (generators were *kaput*) so their only radio link was from the emergency kit. This depended on fast-dying batteries, so they kept telegrams to a minimum.

Eventually, we made arrangements to get a line across joining the ships so we could take the *Colon* in tow. Our chief mate and bos'n assembled some sailors on our foredeck and readied the Lyle gun to fire a line across to the *Colon*. Much to-do as the mate and bos'n ran around the men, ordering this and that ... then ... BANG! went the Lyle gun (primed by gunpowder). This startled the animals congregated (mostly in cages) on the *Colon* foredeck. By the time they'd calmed down, sailors on the *Colon* had grabbed the line, pulled on the chain-link strand it was connected to and made it fast. We passed a stronger chain-link strand and soon the two ships were securely linked. We took off, pulling her along behind us.

The chief mate, bos'n and others were on the main deck, anxiously watching. Eyes were peering from both bridges and from portholes, doors, 'tween-decks -- everywhere, on both ships. I had a very fine view of the operations from the radio shack, way up high, through a porthole looking aft. The *Colon* looked very odd, her deck sloping at a crazy angle, but she was moving forward smoothly. It was dusk and soon would be dark. She only had a few oil lamps for light, but otherwise all seemed OK.

By 7 PM I had finally made a good two-way contact with Miami. We got the Old Man gabbing to Captain Bryant at his home in New Jersey. He was the top man in Grace Line. Company policy, legal requirements, etc., all that stuff was his baby. Even so, Captain Lawson made all the final decisions ... he was the man on the spot in every sense. He'd get most of the praise or blame for whatever happened.

The voice of Captain Bryant trickled through from 2,000 miles away, alternately booming and fading. He told our Old Man that a long, legal-language telegram was coming. Its gist was that a salvage claim had to be made over the VHF, with at least three witnesses, i.e., aural witnesses, from the *Santa Clara* personnel (officers or crewmembers).

This claim would be made on behalf of Grace Line plus the Master, officers and crew of the *Santa Clara*. It would have a "no cure no pay" clause, meaning just what it said. If we couldn't get the *Colon* safely to port, they'd owe us no money. If we did, then oodles of it should be coming to us via Lloyds of London, with whom most ships in the world are insured. Having said which, Captain Bryant's voice, static-crushed, faded out.

"No matter," said the Old Man. "I've got the general picture. Try to get that confirming legal message."

He looked through the porthole in the radio shack. Through it he and I had an excellent view of the *Colon* in tow. John Lawson, Master, was looking at one of the strangest sights of his seagoing career.

He was puffing, as usual, on his pipe.

"Hmmm." (His pipe spurted smoke.) "Have you spoken to their radio-man recently?"

"I've been waiting for him to call, Cap'n. He doesn't want to use up the only battery they have, so he doesn't talk much. I keep that receiver over there tuned to 480 khz. He can call there any time and get my attention."

"We-e-e-ll (he yawned), as long as everything's under control, I'm off to bed. Call me if you want me, anytime. And please try to get that legal message."

Captain Lawson went below to his cabin. I peered through the porthole at the drunkenly leaning *Colon*. I couldn't see the animals; they were on the foredeck. The light was dim, and there were no lights on the ship except for a few dim oil lamps and flashlights. Soon the stars would be out, but no moon tonight. I thought about it. Newsworthy, to say the least. I decided to alert my wife in New York. She could contact the *New York Times*. They might be interested in a spectacular salvage job. Then there was the *Daily News*; they had readers who were bleeding hearts for animals. I contacted WCC -- Chatham, Mass., on Cape Cod -- the biggest, busiest U.S. Coast station for CW operation with ships worldwide.

> WE TOW INTO BUENAVENTURA COLOMBIA PACIFIC SIDE SMALL BRITISH SHIP WHICH BROKE DOWN ITS CARGO IS A CIRCUS THERE ARE ELEPHANTS TIGERS ZEBRAS HORSES ALL VISIBLE ON SLOPING DECK STOP LLOYDS OF LONDON GETTING SALVAGE CLAIM STOP CALL NEW YORK TIMES TELL THEM GET FULL STORY FROM GRACE LINE TELL DAILY NEWS TOO THEY MIGHT ARRANGE PLANE GET PICTURES OF ANIMALS ON DECK STOP ARRIVE BUENAVENTURA SUNDAY.

Well, neither paper took the bait. If the ship had been nearer, and heading for New York, the story would have appealed to them; but it must have seemed so remote -- 2,000 miles away, in another ocean. To me, working in and with radio waves, this distance was only a forty-sixth part of a second in time. Viewpoint governs attitude. Of course, thousands of miles would mean little to a newspaper if the subject was an earthquake or political explosions. Get excited over one radiogram about something bizarre? Huh-uh. Nothin' doin'.

I got the expected Grace Line message from WSL -- Amagansett Radio, New York -- the second busiest CW station working ships.

> REFERENCE SALVAGE SERVICES TO BE RENDERED OR NOW BEING RENDERED

TO THE SHIP COLON ARRANGE AGREEMENT MASTER COLON ON BEHALF MASTER OFFICERS AND CREW SANTA CLARA THAT COMPENSATION IN FORM OF SALVAGE AGREEMENT NO CURE NO PAY WITH SECURITY TO BE FURNISHED LLOYDS COMMITTEE AT LONDON FOR ARBITRATION THERE STOP SECURE THIS AGREEMENT IMMEDIATELY BY WITNESSED VHF COMMUNICATION CONFIRM ABOVE UNDERSTOOD AND ARRANGED.

GRACE LINE

The Old Man tightened his mouth as he read this.

"It says 'immediately,' Sparks, so we'd better get with it. There's you and I. The chief mate will be along in a few minutes with an AB. That's four of us. How are we for radio communication?"

"A bit shaky, Cap'n. It's fine, as always, on 480 khz. But the mates say VHF contact comes and goes all the time. I know it sounds crazy -- there they are, only yards away -- but their voices fade away or disappear in wild static. They haven't got a good VHF transmitter or a decent walkie-talkie."

"Well, how do we sound to them?"

"Fine, as far as I know. Had no complaints."

"That's all that matters. I'll give their captain the dope, and he only has to signify 'yes' or 'no'. If he has any comments we can't get, we'll get 'em on CW."

That's just the way it happened. The invisible Captain Weiner seemed very cooperative, and signified his understanding and acceptance of the salvage agreement by a gruff "Yah, OK" every so often. He thanked Captain Lawson for his help several times. The towing job was proceeding smoothly. A sailor back aft was in contact with the bridge to tell them if anything went wrong there. The 12-degree list of the *Colon* hadn't grown worse. There must have been some discomfort on that ship but we heard no complaints.

It was 8 bells (midnight), time for a watch-change, time to get to bed after a tiring day. I told the second mate to call me if necessary, checked that the auto alarm was chattering quietly, as it should, maintaining the 500 khz watch, and rolled into my bunk.

Dawn came. The sea was calm. Both ships were rolling along smoothly in tandem, only three degrees above the Equator. A brassy sun vibrated in a cloudless sky. No radio station even gave a weather forecast for this area. It was a meteorological no-man's land; never any changes to bother about. We sailed peacefully through the morning.

The sun would reach its meridian at 12:35 PM local time. Navigationally speaking, that would be the real noon -- when the second mate strode on to the wing of the bridge to aim his sextant at the sun. But half an hour before this (when all the clocks on the ship had pointed to 12), bells had rung, watches had changed and the ship's horn had blared noisily. All that fuss on any ship is the dividing point of the day.

After this, on ships, the general mood has always been relaxed: have your lunch, flop down, take a siesta. As Noel Coward said, "Only mad dogs and Englishmen go out in the midday sun." Those who chose to live and work in the engine room will have to put up with stifling heat under electric lights.

Curtains were drawn across portholes everywhere. Just about everyone aboard was sleeping or reading. The shattering of this peaceful scene took only one word. It is possibly the most terrible word to all who go to sea. FIRE!

At 8 minutes past 1 PM, I was roused from sleep by a squawking from the receiver which had been set to 480 khz. The *Colon* had tried calling the mate on watch via the VHF. Whatever the cause, though -- atmospherics in the sizzling heat or weak batteries -- they couldn't get his attention. So, it was back to the good old CW.

> MASTER SANTA CLARA URGENT FIRE IN ENGINE ROOM PLEASE SEND FIRE EXTINGUISHERS WE ARE NEARLY OUT OF THEM SOONEST PLEASE URGENT
>
> MASTER COLON

I didn't hesitate to shatter the Old Man's siesta. Soon everyone else aboard was alert, too. Bells were ringing all over the ship--phone bells, alarm bells -- in engine room, steward's department, crew's mess -- getting all personnel to action stations. The engine-room telegraph in the wheelhouse was swung to STOP ENGINES and slowly the *Santa Clara* rolled to a standstill. The tow line linked to the Colon went slack. By that time plenty of things were in progress, and Captain Lawson told Captain Weiner:

> LAUNCHING LIFEBOAT TO COME TO YOU WITH PLENTY FIRE EXTINGUISHERS SOONEST.

The chief mate and several ABs got busy at number two lifeboat. The bos'n and his men searched the ship for fire extinguishers and brought them to the lifeboat. The passengers, roused from sleep by the hullabaloo, were pleased to be getting more than their money's worth out of this trip by the unexpected events. Some had brought their cameras, which reminded me that I had a movie camera, forgotten in the excitement. I got it.

The newspapers had their opportunity, said I to myself. At least, I'll get some record of all this.

The lifeboat swung out on its davits. The *Colon* had drifted away, swung around, and now was abeam of the *Santa Clara*. I had a very good view of the animals and enthusiastically kept my movie camera on them.

Another camera-worthy scene was on the foredeck of my ship, where our young third mate, pleased with this cue for authoritative action, was organizing sailors and fire-fighting gear. Fire hoses were uncoiled. Foam guns, mounted like machine guns, were ready to gush forth foam or water as needed. Even so, we knew that the fire on the *Colon* was down below, and if it spread upward to the main deck, probably nothing could save the ship.

I got shots of lifeboat number two as it wobbled out on the port side and as it descended into the water. The chief mate, several ABs and dozens of fire extinguishers were on it. I wondered what could be done if the fire got out of control. Not much trouble to transfer all the human beings from the *Colon*, but the animals might give problems. The winches could hoist the elephants if the beasts could be kept quiet while straps were put around their bellies. The horses and zebras were smaller but nervous and frisky. Lifting the cages holding the tigers would be a cinch .... My rambling thoughts were interrupted by signals coming out of the shack...

WGLP DE ZDKM QRK? (How do you read me?)

I dashed through the door, switched on the transmitter. It would need 35 seconds to warm up. Of course, this time period seemed like two minutes when my dominant mood was anxiety. CLICK! The transmitter was alive, ready for operation. I hit the key:

ZDKM DE WGLP R K (Get you. Carry on.)

The Colon R/O cut out the usual preamble signals:

REQUEST TRANSFER WOMEN AND CHILDREN TO YOUR SHIP FOR SAFETY

MASTER COLON

"Forgot they had a circus troupe -- or whatever you call 'em," was the Old Man's reaction. He had been standing on the port side of the bridge, watching the lifeboat pull away.

"We'd better contact the chief mate by walkie-talkie first," he added.

This was done, then a CW message was sent to the *Colon*:

ZDKM DE WLGP MASTER COLON YES OF COURSE STOP CHIEF MATE INSTRUCTED TO OFFLOAD WOMEN AND CHILDREN

MASTER SANTA CLARA

"You'd better see if you can get hold of Miami, Sparks. Captain Bryant has to be updated."

He got more than that. He also got a sort of running commentary on events as they were happening. Atmospheric conditions were excellent. I even arranged another schedule with WOM, to update Bryant again, but it was not to be. An electric storm fouled things up. Speech was unintelligible. The SSB (single-sideband voice machine) just couldn't cope. So, once more, back to reliable CW. WOE -- the Miami Morse station -- was soon contacted and ready to receive.

I reflected on how often in this situation CW had come to the rescue. Voice transmissions, whether on the sophisticated SSB, the comparatively simple VHF, or even the much simpler walkie-talkie -- all had been unreliable. Atmospheric conditions ruled everything. Yet a simple sequence of dits and dahs had wriggled or punched through every time. Years from now, when an Old Man can communicate with his company office from 5,000 miles away via a two-way TV/telephone linkage (seeing and hearing perfectly) -- there will still be a need for a human hand on a Morse key. In emergencies they will be necessary. There is no foreseeable end to CW, with its dits and dahs.

Something else always in favor of CW -- it forces the sender to collect his thoughts and summarize them into "cablese." Radiophones allow tongue, brain and emotions all the freedom they want -- too much -- to muddy things up. Clarification and repeats waste more time. After all the excitement and uncertainty, it was comforting to have stuff boiled down to a few words:

> GRACELINE NEW YORK ATTENTION CAPTAIN BRYANT FIRE ON COLON STOP COMPLETELY EXTINGUISHED BUT ENGINE ROOM FLOODED TO OVER FLOOR PLATES STOP NO INJURIES BUT NO STEAM NO POWER COLON IS DEAD HULK STOP LIST TO STARBOARD NOW 15 DEGREES BUT HAVE COLON AGAIN IN TOW OK STOP PROCEEDING BUENAVENTURA 5 KNOTS STOP SUPPLIED MANY FIRE EXTINGUISHERS AND REMOVED WOMEN AND CHILDREN AT MASTERS REQUEST THEY ARE NOW COMFORTABLE IN SPARE PASSENGER CABINS
>
> LAWSON

There were 12 passenger cabins. (Above that number the ship would have to be reclassified as a passenger ship and a 24-hour human watch would be kept in the radio department -- needing at least three R/Os.)

Our six passengers had the center of the ship to themselves -- very comfortable, thank you: swimming pool, fully-equipped bar and even a grand piano. Into this luxurious atmosphere came the refugees from the *Colon* -- women and kids who were ragged, barefoot gypsies, colorfully dressed, speaking only Spanish.

Someone rapped on the door of the radio shack.

"Come in," I called, but no one did.

I opened up. There stood a beautiful young woman in a gypsy outfit of violent colors.

"*Telegramma, por favor*?" was her shy, husky query.

"Oh, er, yes. *Si ... si* -- come in -- *entrada*!" I stammered.

Barefoot, she snaked her way in, hundreds of beads jangling around her. She already had her *telegramma* written out in block-lettered Spanish, and had some American dollars, so business was brief.

"*Gracias*!" she smiled, and slunk away.

"Wow," I muttered to myself as I set up the radio gear to call OBS3/Callao (Peru). Her pungent perfume still lingered, as did the memory of her voluptuous movements and her luscious little partly exposed breasts.

My Spanish is next to nil, but she seemed to have been begging some circus HQ to come urgently to Buenaventura as they were in difficulties:

> CIRCO EGRED LAMIURA (PERU) VENGAN URGENTA A BUENAVENTURA EN DIFI-CULTADES POR MARCHA LO LLEVAN A REMOLQUE
>
> ROCA

I later had slinky Roca's message translated: The EGRED circus folks in Lamiura were indeed being asked to come urgently to Buenaventura -- because "that's where we're being towed to."

Whether this request brought results, I never found out, but this contact with Callao Radio was useful. They told me they had a message for the *Colon*, which I naturally relayed:

> ADAM WEINER C/O MASTER COLON VIA WLGP URGENT YOU ADVISE IF YOU GOT

TUG TO BUENAVENTURA STOP IF YOU CANNOT CONSIDER RETURNING TO CANAL FROM BUENAVENTURA WHY NOT SHOW CIRCUS THERE WHILE AWAITING PARTS AND REPAIRS

WILLIE

The *Colon* had been drifting toward a rocky coast and inevitable shipwreck. She had had a fire aboard twice. Many lives had been in jeopardy. We were now performing the basic job of a tug. But businessman Willie had only one concern -- THE SHOW MUST GO ON. To be fair to him, perhaps he had very little idea of what had happened.

In contrast, the alarm in New York (at Grace HQ) was excessive. They pictured the *Colon* being slowly towed along, in dire danger for several reasons; her engine room was flooded, and she had no power of any kind; she was heeling over at a crazy angle, therefore liable to capsize if a large wave hit her side. Perhaps she'd catch fire again at any moment? All these fears seemed to justify:

MASTER SANTA CLARA REPORT CONDITION EVERY THREE HOURS OR SOONER IF NECESSARY PHONE PORT CAPTAIN 516-427-6743 STOP KEEP AMVER ADVISED IN CONDITION OF TOW

GRACELINE

They're right, I thought. I'd better tell AMVER. It's not mandatory, and there's no emergency situation, as Grace thinks, but we'd better let 'em know about everything. "Automated Mutual Assistance Vessel Rescue" System is worked by a computer in New York City, into which vessels of all nations feed info. They give their position, destination, route, etc., and from that info, when anyone gets into trouble, the computer figures out which vessels are nearest and able to give help. In our case, assistance was already being given; but AMVER would be interested.

"I'm getting the mates to work out an AMVER message," said the Old Man (fussing with his eternal pipe), "but get the company on that phone number if you can. I'd better pour some oil on their troubled waters! They don't realize the sea around us is like a duck pond, with no bad weather in the offing."

Another telegram came from the worry-warts in New York:

> MASTER SANTA CLARA WOE/MIAMI REQUEST MASTER COLON ADVISE YOU HIS ESTIMATE SERIOUSNESS CONDITION UNDER PRESENT CIRCUMSTANCES AND IF HE CONSIDERS SERVICES NEAREST TUG WITH PUMPING EQUIPMENT SHOULD BE REQUESTED STOP PRESUME MASTER COLON HAS ADVISED HIS OWNERS THROUGH YOUR RADIO IF NOT ADVISE HE DO SO STOP CONFIRM FOREGOING WITH YOUR ESTIMATE SITUATION
>
> SMITH, GRACELINE

"Dear old Smitty," chuckled the Old Man, "clucking away like a mother hen. We should make Buenaventura pilot station around noon tomorrow."

We did. We pulled the *Colon* past the breakwater into Buenaventura anchorage at 12:30 PM next day. The tow line was retrieved. The Colon anchored. Officials, agents, repairmen boarded her. Mission accomplished.

Then came an ugly twist. Captain Weiner (probably acting on orders from his agent) reneged on his agreement. He seemed a nice guy and was sincerely grateful for the help we'd given, but his advisers (maybe including anxious Willie) probably figured out that our having only aural witnesses was a weak point. Eyewitnesses or a written agreement--both were missing. Anyway, Weiner wouldn't sign anything.

Captain Lawson, acting on orders from New York, put a lien on the *Colon*. This made him -- as his officers and crew kidded him -- the new Circus Master.

"We have to formalize a complaint, register the lien, and all that jazz, before a U.S. consul," the Old Man explained to me. "We won't be able to do that in Callao, our next port, since we arrive and depart on a weekend. We'll have to wait until we get to Valparaiso."

Valparaiso (Chile) is as far south of the Equator as New York is north of it. Even so, locked between the Andes and the Humboldt Current (coming up from the South Pole), it is always cool. The harbor is full of strange birds. Somehow they know this is a bird sanctuary.

Into the U.S. consul's office trooped Captain Lawson, the chief mate, the AB John (he had a knack of always getting into the picture) and I. There we started the process that was to be continued in the London office of Lloyds and elsewhere.

As the *Santa Clara* sailed north, the officers and crew had a permanent conversation piece -- the salvage award. How much could be expected? A million dollars was the favorite figure to start from, before it was divided up between company, master and crew.

* * * * *

Let's go back to that deck chair where I was stargazing and ruminating about time, light-year distances, historical perspective and all that. Let's escape from that hustle-bustle mood of pragmatic decisions about the Colon.

See that fuzzy little cloud of stars up there? It's the constellation Andromeda--2.3 million light years distant (13.9 million million million miles away). To an astronomer, though, it's close--the next nearest galaxy to ours--after the Magellanic Cloud, south of the Equator, which is a mere 1.7 million light-years away. You know what?

**everything depends on
your point of view**

Homespun philosophy. It antedates Albert Einstein, whose $e=mc^2$ *is basically just that. So, for change of viewpoint and pace, let the machinery of The Law take charge. It's S-L-O-W.*

A year went by; no word from Grace about the salvage claim. Two years...three years...four.... "No, we haven't forgotten," they told me. "Don't be impatient."

Five years -- nothing. Six years -- and Grace Line merged with Prudential Insurance, Pru being the boss. They had to; the alternative was bankruptcy. The shipping line was renamed "Prudential Grace," then simply "Prudential." Poor W. R. Grace (who founded the Line about a hundred years ago) must have rolled in his grave. The ships got new funnels, new everything.

Fortunately, in all the reshuffling, the obligations of Grace Line to its employees were not forgotten. It was seven years after the Floating Circus incident when I got my share of the booty for helping in the rescue...

$90.00

That was that....end of story....or was it?

"Time marches on." (Maybe it just trickles by....) Anyway, I have the annoying habit of not being tidy and leaving things alone. I fidget.

P.S. Much later -- it's too embarrassing to put a date down -- I wrote to Lloyds. I carefully identified myself. Would they kindly let me know what that Final Award had been? They obligingly searched their records. The sum was $4,965.00.

Hmmm. That would be $2,482.50 to Grace Line, $1,241.25 to Captain Lawson, leaving another $1,241.25 to be divided up among the 43 officers and crew ... Yes, $90.00 would just about settle it....

Such is the authority of Lloyds, that I accepted theirs as the last word, until time started trickling by again. Like the proverbial penny, stuck in the machine, but wriggling through at last, I got a long-delayed reaction....

P.P.S. I've been soaring up to the stars and back. I was honestly trying to get a perspective that a sizable hunk of time and space would give. But it would also be a useful device to escape one little question:

Where the hell (any year, any place) can/could one get a complete ship (including hull, fuel, machinery, etc.) Plus a complete circus (elephants, tigers, the works) FOR $4,965.00????????

(no more...FINIS!)

CHIRRUP OF HAPPINESS

The 20th century crumbles away as you leave Canal Street in New Orleans to enter the French Quarter. The clip-clop of horse-drawn carriages, with their jingling bells, are 19th-century sounds to offset the booming horns of freighters and tankers as they head for Southwest Pass and the open sea.

When I'm in New Orleans, I make a beeline (through ancient, gaslit streets) to a relic of the 18th century -- the Napoleon House bar. When Napoleon was first exiled in Elba, there were plans to rescue him and bring him to the New World -- specifically, this spot. Hence, now, on the walls, many Napoleonic pictures and memorabilia.

The drinks are good, the atmosphere is cozy -- and, above all, there is classical music in the background instead of wailing, moaning, rock-and-roll, etc. (If real Dixieland jazz is available, I'm all for it -- yesyesyes -- but that's a few blocks north, on Bourbon and St. Peter.)

Beethoven had admired Napoleon and dedicated his *Eroica* Symphony to him. Later, when Napoleon crowned himself emperor and revealed his megalomania, Ludwig shed him like dirty laundry. That was in 1805. Nevertheless, Beethoven and Napoleon are linked in my memory, and both are linked with the Napoleon House. When they see me coming, they fix me a Brandy Alexander and put aside Beethoven's Seventh to be next on the turntable.

Still spinning the clock hands around and back, let's settle on Rotterdam, 1954.

* * * * *

I was in love with a Dutch nurse. Everything revolved around Nelly. Nelly's pretty golden head, laughing from the other side of the canal (we had parted, each on one side, just before the moving bridge went skyward); Nelly's trim figure, tripping along Koolsingel to meet me; Nelly, when we were out on bicycles in the country, calling over her shoulder, "I'm so happy!"; Nelly, eyes closed in rapture among pillows in the Little Flushing Hotel where we used to come for weekends. (We went once too often. My ship had been a month in Rotterdam, getting repairs. Then she weighed anchor while I was with Nelly in Flushing); Nelly in London (I was between jobs over there, and got her to fly across one weekend) with Trafalgar Square pigeons clustering over her; the changing of the guard just a background to her face; Windsor Castle just another incidental background to her. Oh boy, were we in love. In the Candlelight Restaurant where we had supper she glowed with beauty and loving surrender... even more so in the Regent Palace Hotel, off Picadilly Circus. It wasn't just physical; it was total. Nelly, Nelly, Nelly was my obsession.

I desperately wanted a job that would bring me back to Rotterdam, but I didn't know how to get it. It seemed as though we'd be separated for an indefinite time. First, I joined a company which put me on a ship lying at anchor in the Holy Loch, near Glasgow. Day after day, then week after week, the ugly, rust-covered hulk lay surrounded by the beautiful Scottish countryside, awaiting orders. My lonely occupation was to sit in a cafe near the lake, writing love letters to Nelly. Pot after pot of tea was replenished by the amused waitresses. Thousands of words streamed out of my pen.

I got fed up with waiting. I reregistered with another company. They had a ship in Vancouver needing an R/O. I was flown across to New York and transferred to another plane going to Toronto; thence to Winnipeg and Vancouver. I joined the ship, another filthy Greek ship under the Liberian flag, but I didn't care: I was mentally heading for Rotterdam, but (in fact) I wasn't. The ship was a tramp with an uncertain future. She started crawling down the West Coast at about 9 knots. My misery increased with every slow mile southward. I couldn't visualize how/when/if I'd ever see Nelly again.

I didn't sink quite as low as one of the fireman. He was worried abut his wife having a love affair to pass away the long time he was absent. Plus, they had a gas leak in the engine room and he had to work a lot of overtime because they were shorthanded. I did my best to allay his anxiety about his wife by getting radio messages to and from his brother in Athens. (Contacting that station from *anywhere*, let alone the Pacific, is a hell of a job.) His brother told him (in Greek -- I didn't understand) not to worry. He did, though. The cumulative effect of his other pressures piled up. One day he went to the stern of the ship at lunchtime. He carefully laid down his slippers, side by side, attached a rope to the rail, and lowered himself into the wake. The captain turned the ship back and I sent an XXX (second in importance after SOS) to all ships to look for him. He was never found. (Those slippers, lying side by side, haunted me. They still do.)

This tragic incident epitomized and emphasized the black cloud of depression hanging over the ship and the mental outlook of all aboard, what with gas leaks down below (producing a near-mutiny by some of the crew there), the filthy conditions on the ship and my yearning to be with Nelly again in Rotterdam. Well, by the time we had passed through the Panama Canal, crossed the Atlantic, entered the Mediterranean and anchored in the outer harbor of Genoa to wait for orders, I was *very* miserable.

That night I looked at the program notes from the BBC. There was a concert that evening featuring Beethoven's Seventh Symphony. Now *that* might shake me out of my miserable torpor. It would be on about 8:00 PM. I'd get the 7:00 PM traffic list from GKL (Portishead, the British main station for ships). There would probably be nothing for us, but I had to check. Even if there was a message for the Old Man, I'd have time to get it and still hear Ludwig around 8:00 PM.

The call signs of ships were sent alphabetically. Mine, ELYK, could be expected soon. The E's commenced ... ELAP ... ELAP. They were sent twice. Some ships would be thousands of miles away and hardly able to hear even a once-sent call sign. I was just across Europe from GKL -- in the radio sense we were merely a stone's throw apart, and he was deafening ... ELCY ... ELCY. ELJT ... ELJT ... ELYK (Yes!) ... ELYK ... I switched on the standby switch, and the filament voltage registered immediately. I needed 35 seconds for the transmitter to become operational. GKL still had all the rest of the alphabet to go through, but I could get hold of one of his buddies on another frequency, while the traffic list was still pounding out. I put one hand on the Morse key, watched the DC milliameter at the top of the console ... fiddled for maximum ... got it! ... ELYK was readyready. Called, GKL GKL GKL GKL DE ELYK QRK?

He didn't hear or reply.

I tried again.

This time he stopped, and I hoped he was going to call me. No, the call sign was F: it was a French ship. Then he was silent, listening to the Frenchie or someone else. I looked at the clock. Why not quit and get the message after the concert? The Old Man's message isn't important. I bet it can wait an hour or two, anyway.

While I'd been thinking thus, I'd been pounding the key: GKL GKL GKL GKL GKL GKL DE ELYK GKL GKL GKL GKL.

He stopped again.

I sent my call sign a few times, and halted also.

ELYK (Hurrah -- he was calling) ELYK ELYK DE GKL QSS? (What frequency do you want to send on?)

I told him. He answered that I was number two, after the French ship, so there was a damned good chance of getting everything cleared up before the concert.

The French ship was taking up more time than I'd expected, and I was beginning to glance at the remorseless minute hand of the clock moving towards the 12, when GKL called me. I answered -- various formalities, civilities, etc., then as his Morse came crackling through the earphones, my fingers danced on the typewriter keys, and I watched the message churning off the typewriter roll:

MSG NR 1 LONDON 11 27 1803 GMT
MASTER SS VASSILIOS/ELYK GKL
AFTER DISCHARGE GENOA LOAD
SUFFICIENT BUNKERS FOR ROTTERDAM

GRELCOSHIP

The bulkhead above me -- the sky above that -- both cracked open; colored tinsel and pretty colored balls seemed to be raining down on me! Rotterdam! Rotterdam! I acknowledged the message, shut off the transmitter and danced out of the door onto the open deck. I'd *rush* this up to the Old Man. Rotterdam! The stars were no longer coldly impersonal, but twinkling merrily as stars should. The moon's expression was a faint smile. The lights of Genoa were shimmering. Oh! I was happy!

I got the message to the Old Man, and minutes later I was snug in the radio shack, feet up on one chair, earphones off, three loudspeakers positioned to give a stereophonic effect, pleasurably snug, readying my ears as the various instruments tuned up, and the unseen audience quietly murmured. I'd better make up a joyous telegram for Nelly and get her to share this delirium. A burst of applause signified the conductor entering on the scene, then silence as he held his baton aloft. Quietly, the symphony began. . . .

It was a rumbling start, with promise of great things to come (or was it just hindsight wisdom to say that?). Pretty little tunes kept starting; then dwindling off as if suddenly bashful; reappearing, more confident; instruments ganged together, then advanced together; there was gathering momentum, more delay, then on-on-on together up-up-up...power! Repressed energy leaking out; the trumpets rose up, beckoned all, and the strings delightedly followed.

The second movement is at first somber, determined, slow and steady. Then it gets cheerful. (Like the man who told Dr. Samuel Johnson, "I, too, have tried to be a philosopher. But try as I would, cheerfulness would keep breaking through!") Odd but lovely little turns of melody went diving downwards, wriggling up again through different instruments and different interpretations. Then all the instruments gathered force, section by section, advanced together. They were resolute, determined--on, on. Forward, forward, then back...change key; change pace. On, on, on again. Long pull...drawn out to a big humorous question mark.

No hesitation for the third movement. Here was the answer to any query -- savage impetuosity! Volcanic energy! Pounding relentless rhythms with the cascades of beautiful sound! The very first audience at the University of Vienna's great hall, on December 8, 1813, was so overcome by the demonic frenzy that they sat silent for a moment at the end of the movement, then rose and hollered for an encore! (Unthinkable! Middle of the symphony! Never been done before!) Beethoven himself was conducting. He blinked at the exuberant audience, shrugged, and gave 'em what they wanted. The enthusiasm of audiences has hardly diminished since. "He was drunk when he wrote this symphony," cried Clara Schumann's father. "It is the apotheosis of the dance," said Wagner. A hundred and sixty-three years after that first performance in Austria, I heard an audience member spontaneously yell, "Hurrah for Beethoven!"

For me, though, the apex of excitement came at the beginning of the fourth movement, just after the pent-up fury of the third movement had been abruptly halted, then left on a shelf, as it were, to come cascading down again as soon as the fourth movement started. There are three little notes that have little or no importance in themselves. They're grace notes, throw-aways, silly little things that just happen to be there. No, that's not fair. They wouldn't have such effect on me if it hadn't been for that relentless driving third movement. All of that is implicit in those three little tripping notes.

They have syncopation, a little offbeat, and I like conductors to make them incisive, not letting the orchestra slide over them sloppily. Erich Leinsdorf manages this best... DIT. DIT. *DAAAAAH*! When I was deliriously happy, in Genoa's outer harbor, those sharply clipped notes had another message -- Rotter-*dam*!

Their broken offbeat rhythm seemed almost like a hiccup. I dubbed them "hiccup of happiness." Then I thought "chirrup of happiness" was more apt. Hiccup or chirrup, they had and have an effect out of all proportion to their significance/timing/position or anything else at all.

Up with them, on the same plateau for extraordinary effect from tiny beginnings, I place five little grace notes in the song "Someone Will Say We're in Love" (*Oklahoma*!). Decorative embellishments -- that's all -- But oh! Delightful!

Well, that's where I'd like to end this piece. The chirrup (or, if you will, the *hiccup*) of happiness, a mystifying title, has been explained.

"Ah, yes, but what happened to Nelly?"

Well the ship went to Rotterdam, as scheduled. Nelly and I made a date on the phone to meet in our usual spot -- Marconiplein, a square near our favorite bar. We were both there about the same time, ran towards each other and collided, in bliss, right in the middle. Can I end the story there?

"Certainly not, if there's more to tell."

Life isn't neat. Art is. One can start a story wherever one likes, end it when some point has been made. In real life, though, all the past is a prologue. Something might be ending, but something else is starting.

"You're hedging. The love story didn't lead to wedding bells. You broke up, yes?"

Well, yes, but...

"Ah."

Don't be smug. It's a temporary unhappy ending. I'm certainly not going to open the can of worms that explains how/why we broke up. We went our separate ways...and we both got happily married. There. Does that satisfy the urge for neat endings? Children's stories are tidy. They end with "and they lived happily ever after." Which we know doesn't always follow. Another story

BEETHOVEN
7th Symphony, 4th Movement

then starts. My point is that the Nelly love story, against a background of Beethoven's Seventh (pouring over me, on a ship lying at anchor in Genoa Bay), should be allowed to stand as is, without follow-ons, complications, cans of worms, etc. My particular interest, not just in the perpetually acclaimed third movement, but three silly little notes that follow it, at the start of the last movement--well, that's something else again. Hiccup. Chirrup. Trip-up. Call 'em what you will. They've got no right to make me flip, but they do.

A hundred years from now, a conductor will hold his baton aloft as the orchestra beneath him gets ready for the B. #7. It will start fairly quietly. There will be members of that audience who have never heard it before. They may join the huge throng who have listened to it since December 8th, 1813, and either react quietly, or go a little crazy with delight.

SUMIKO

Should I be ashamed of what happened? Or can I plead ignorance and even be a little proud of a quality that can be called either pigheaded obstinacy or praiseworthy persistence, depending on your angle of vision? You judge. The story -- the girl -- has been going round and round my head for years. You'll get it all in a few minutes' reading time.

* * * * *

どうぞ  "DO-zo!" ("please")

ありがとう  "Ari-gaTO!" ("thank you!")

These two most commonly heard words in forever-polished-and-formal Japan were all around me in the Tokyo street. It was nice to be back again, after so many years.

I was taking a routine visitors' tour. Our female guide was explaining -- sometimes with shy apologetic titters and always in amusingly quaint English -- what she thought we should know. Behind and under all her explanations, historical anecdotes and so forth the everyday facts of Japanese life came through: the shuffling, kimono-clad women (only about half -- the rest were in short, modern skirts), the traffic cops, gesticulating and whistle-blowing from cubicles high over everyone, the strange and beautiful calligraphy (just ordinary printed characters to the Japanese, but to the foreign visitors these ideograms were almost paintings, with thicks-and-thins and long brushstrokes) and, through it all the constant exchange of those musical words *"DOzo!" "ArigaTO!"* (the last syllable, whether accented or not, going up).

This sketchy background is to offset my personal memories. For me, all things Japanese revolve around a shyly tittering little girl, Sumiko.

How well I remember my first visit to Japan, in 1955. The ship's horn blasted as we entered the tiny port, as a way of saying "We're here!" Who could hear or care? It was an hour before midnight. There were no lights, no ships to be seen. My ship was an oil tanker. Tankers often go to out-of-the-way places, but this time the proverbial "back of beyond" seemed to have been reached. I only knew it was some small port in Japan, far away from big places like Yokohama.

To me this ship represented both a triumph and a retreat. Triumph, because it was the result of my writing about a hundred letters to American shipping companies, looking for a better job with more pay. It was also a retreat, because my romance with a Dutch nurse had collapsed and I wanted to get far away, have a complete change. When the New York company founded by Emil Ludwig offered me the R/O job on the *Bulkpetrol*, I grabbed it. Ludwig was the American tycoon who rivaled Greeks like Onassis in ruthless empire-building with ships of Liberian registry -- a "flag of convenience." His company saved money by ignoring certain safety standards, etc., but they offered me good money. They contacted me in Rotterdam, then raced me across Europe in a train to join the ship in Naples.

I had been very much in love with Nelly, a Dutch nurse. The trauma of separation left me with a very crude need. I was woman-hungry. Centerfolds from *Playboy* and pinups from other girlie mags were taped all over my narrow cabin and cramped radio shack.

Since this was a very big ship -- for a time she was the largest tanker in the world -- these remarks about "narrow," "cramped" need expanding. Ludwig was a multimillionaire with the idiosyncrasy the very rich often have: to economize on relatively unimportant things. For instance, instead of strong solid masts to hold the ship's antennas aloft, he substituted skinny wooden poles. I thought, "just wait until a strong gale blows them away. His false economy will boomerang." Another instance was his crowding the deck officers, captain and radio officer into the cramped superstructure, for'd. You couldn't swing a mouse, let alone a cat, where the navigators and I had to live and work.

Is all this irrelevant and uninteresting? It was no small thing to us, then. It popped up only because I was describing the background for girlie pinups. Back to that subject; in an all-male world it can become an obsession.

Aboard the *Bulkpetrol* it certainly was that. We were at sea for such long periods, ashore so briefly. Then we hit the beach in San Pedro, California. All the centerfold bosoms and legs faded before the shimmering prospect of real, three-dimensional beauties in shops, on the streets, everywhere -- visible, tangible (oh! to touch them!). San Pedro is a suburb of Los Angeles. Into that area for many years -- since the movie industry began there -- have flocked beauteous/pretty/devastating/contest-winning girls, would-be movie stars. One innocent (?) little girl, waiting on me in a diner, excited me to fever-pitch. "Compact" was the word vibrating incessantly in my mind as her rounded arms, legs and breasts hovered over me with plates and coffee. She was cool to my advances -- and I was easily deterred.

Still, she was only one of many, many females. They were all lures, taunting me with an overriding thought -- "gotta make the most of my time before I go back on the ship." Then came a rebel thought -- "Dammit, why go back at all? I won't be needed. The ship's only going out to anchor, to clean tanks, then returning to port. I'll play hooky." I did.

I was nervous. I out-shouted my wobbly thoughts by a Plan: to pick up a woman at a dance hall. This was a place where one could have a dance with any girl one fancied, for a dollar a dance. I was all het up, but jittery.

I messed up everything. I did not "get me a dame," though at one point I had one smiling her come-on to me. In my nervous uncertainty I made a balls-up of the whole situation. As I said, I was even being encouraged at one point, but I couldn't follow through. I can't even remember the details. The body heals itself with a scar, then usually that, too, disappears. The mind has the same technique.

The Old Man, normally very easygoing, was furious. The ship couldn't legally go out to sea (even only a mile or two, for tank-cleaning) without a qualified radio officer aboard. The captain happened to be one! Captain Jorgensen is the only one I've ever met or heard of, who graduated -- well, switched -- from radio operating to become a deck officer and eventually a captain. So the legal requirements had been met. But that, to Captain Jorgensen's exceptional discomfort, was certainly not all. He had to tune in the coast station, KFS; get a message from them, and send one. It had been many years since he'd touched a live key; the gear was so utterly strange -- the things to do, the buttons to push, the time to wait, the sequence of operational tricks, the purpose of this gimmick, the result of turning that knob -- he felt like a panicky baby trying to swim. He'd seen me umpteen times, pushing this knob up, swinging that dial around, waiting for a meter to reach a certain point, looking at the clock to allow 30 seconds to pass after pushing that knob before I pushed that one -- but it was a bundle of tricks to which he paid little attention. It was as though he'd been watching a movie with sleepy attention, then -- WHAM! -- suddenly being thrust into the events on the screen and finding them bright, noisy, moving realities all around him.

We had greater empathy than he knew. I remembered standing aghast at the sight of that radio shack with 121 knobs to pull or twist, while the ship went clear across the Pacific Ocean--and I without even five minutes' experience to draw on. Oh yes, Cap'n Jorgensen, I know about the cold void where a stomach used to be. He was in no mood to swap memories or to philosophize, though; he was HOPPIN' MAD. He was still shaky as he bawled me out in his office, with a large whiskey at his elbow.

The result of this encounter was that yours truly was squashed down even deeper into the Slough of Despond. I had no girlfriend. I had no sex object. I'd made a damned fool of myself trying to pick up a gal in L.A. Tankeritis was a grim state of mind and physical restriction. To complain of claustrophobia on a ship in the vastness of the Pacific Ocean might seem ridiculous, but the mental restrictions, being surrounded by rigorous schedules, comings and goings, ship routine so boring, boring -- and the cramped quarters, the narrow, crowded radio shack, with little or no hope of circulating in the civilized world of shops, moving, parks, interesting people and places, things to do, newspapers to tell me about all the world ... and, above all, girls, girls, girls....

Day after day after day, while seagulls wheeled silently overhead then maybe swept down to pick up some garbage thrown over the ship's side ... yes, I know my life in the radio shack was based on the international community on 500 khz and all day, all night, there'd be chattering on this frequency between ships which would be from 500 to 1000 miles away from each other. It would be logical for me *not* to feel isolated, but a mood or feeling doesn't follow reason. I was FED UP, blue, utterly miserable.

* * * * *

This was the background, physical and mental, as the ship hooted to nobody in particular while slipping into this lonely Japanese port, at 11 PM.

Somehow we got somebody to get us alongside, plus some minor official to act for customs/immigration/port and everything. In no time at all I was going down the gangway with some other crewmembers. It was very dark. It was very quiet. We were thinking how idiotic it was to saunter ashore at all at this time of night in such an out-of-the-way place. We went through a big gate *--and giggling girls swooped on us from every direction!*

I grabbed a pretty little one who called herself Sumiko. It might have been anyone. In the daylight, under different circs, I might have delayed my choice, but Sumiko came to me out of the dark, giggling, teasing, a shapely little bundle of fun. Like a drowning man seizing a lifebuoy thrown to him, I seized her, crushed her to me. Her soft body was going to swamp and dissolve the ugly memories, lonely hours and miseries of many months.

When half a dozen crewmembers and two officers beside me had each attached himself to a girl, we all went our separate ways. The girls weren't big-city whores but small-town, good-time gals. Ships came so rarely to this remote village and there was so little contact with the big cities like Tokyo and Nagoya that the folk here had their own laws, habits, outlook. No foreigners had been around for about four months prior to our arrival. It might be many months after we departed before another ship came. So, naturally, they were going to make the most of our visit. Every minute counted. News about our coming had been excitedly discussed, and when we did at last arrive, the man-hungry women inhabitants didn't give a damn about it getting on toward midnight -- they *had* to get down to the docks and give us woman-hungry creatures a big "WELCOME!"

That explains why our confrontation with the eager girls at the gate was unique. Now that Sumiko has come on the scene, I might be expected to concentrate on her, but to understand her we have to look at her friends and relatives, the little community of which she was a part.

No, on second thought -- not yet. Let's take a birdseye view of Japan and the whole situation then and there, ten years after WWII.

It's only about a century since the Japanese came out of isolation, and other peoples have been alternately fascinated/

amused/disgusted/horrified/admiring with respect to them since then. Madame Butterfly and Lieutenant Pinkerton were precursors of Sumiko and me! The shoddy junk of "made in Japan" pre-war stuff indicated the economic trouble they were in, which the warmongers then had thought they could escape by violence; but they got true world conquest via technology in cars, radios, watches, etc.

In all this switching of attention, changing of face and varying focus of interest, the Japanese have had some constants. The emperor and the ancient way of life he stood for were the lodestars for thousands of years; almost the only constant after WWII defeat.

The Japanese have intense loyalties to their families, to their particular neighborhood, (city, town, village, whatever) and, by logical progression, even to the company they work for, be it large or small. That's partly why Japanese cars, radios, cameras and binoculars have become world leaders. Each company is a world in itself, a super-family, giving loving attention to each member from the cradle to the grave. No individual works for himself and his personal family alone; he/she is part of the great company-family, knowing that he/she is securely, lovingly protected by The Company. He/she doesn't want excessive wages. He/she just doesn't worry about anything.

Sumiko was an orphan, but the whole village was her substitute for parents. She was loved and welcomed anywhere. Perhaps it's not even necessary to cite the Japanese temperament and way of life. The mere fact that the whole town was only a few hundred people made it like small towns or villages anywhere in the world. Crime was minimal, family-feeling paramount, everyone was on a first-name basis.

So that was the woman; these were the people, this was the little town I'd stumbled into late one night -- to be around for how long? I'd had eight months so far of these ultra-brief visits to ports ... but oh, joy! Hallelujah! The ship had engine problems. We were going to be here for a week or 10 days!

Happy days and nights with Sumiko. We made love everywhere at all times. She was a popular, fun-loving little person and welcome at the houses of relatives or friends almost anywhere. At each place we took our shoes off at the entrance, crept over the mats in the flimsy structures to a partitioned-off portion you might call a room, and got cozy there. Copulation was spontaneous, natural, accepted. There were no social signs of shock, surprise or disapproval. On the contrary, smiling heads nodded and the eternal chorus of the musical words *"DOzo?" "ArigaTO!"* was the linking theme of politeness where fucking was just an integral part of the chain of orderly civilized behavior. In this mostly agricultural community sexual intercourse was taken for granted. For me there was a very apt parallel with the mutineers of the ship *Bounty*. They too were merchant seamen. They too had had long miserable days at sea. They too reveled in the easygoing, woman-filled shore life. The comparisons end there.

Although it was country life, with an openness about sexual matters, Sumiko must have raised more than a little amusement when her enthusiasm spilled over into orgasmic cries of delight that went through walls of paper!

I thought (as Fletcher Christian and his fellow mutineers had thought, way back in 1790), I want this to go on and on. To hell with the ship. This thought wouldn't, couldn't go so far as desertion but did come to the statement I gave Sumiko again and again, *"I'm coming back!"*

The dreamlike Shangri-La episode had to end. No more of Sumiko's bliss-filled Japanese eyes looking up at me from the pillows. No more interesting vegetables to be balanced at the end of chopsticks. No more hot sake in little stone jugs. No more of the strange, interesting, colorful little habits, expressions and customs that Japan had evolved through thousands of years as a remote island. No more forever-recurrent *"DOzo?"*
"ArigaTO!"

This little please-and-thank-you chant had knitted things together, and I vowed, with General MacArthur, "I shall return!"

On the journey back across the Pacific, I planned to switch to another ship that would come to Japan. A logical starting place for this would be the West Coast, but I ended my year-long stint on the *Bulkpetrol* in New York, after going through the Panama Canal. It was going to be very hard to find a ship on the East Coast bound for the Orient, let alone one heading for Japan. Even if I did find one, it'd probably be just another tanker, in port for hours only. It would be against all odds even to hope for a repeat of the *Bulkpetrol*'s prolonged stay. Jobs to go anywhere were few. I telephoned. I wrote. I visited shipping offices. I moaned and bitched and asked -- seamen's clubs, officials, unions, anybody, everybody. I got nowhere. Not a smidgen of a hope.

My money was rapidly going. I couldn't afford to make the 3,000 mile trip to California to be in a better position to grab a Japan-bound ship. I was considering going west as a hobo on trains. I was still a British citizen and couldn't legally get a U.S. job, so I got one illegally. I started nerving myself to go to the rail freight yards, and prepare to hitch the rails as a hobo. I felt like doing anything short of a crime to get a freighter -- it had to be that -- going to Japan.

All this while, several months, I'd stayed in contact with Sumiko, telling her about my futile struggles. She had no command of written English at all, so we depended on one of her fellow villagers who had; this rather cramped the style of passionate love letters! But my real hopes were getting thinner, smaller, weaker. I was just being obstinate and letting a mechanical routine of inquiries take the place of sincere hardworking efforts (with hope) to get that ship. If I, in the bustling city of New York, couldn't maneuver this simple task of getting a ship, how could Sumiko, in a backwater of Japan, understand my problems, or get the atmosphere of the Big City tempo?

Regarding the latter, I couldn't be nearer to the center of action -- downtown New York, near Wall Street. Like a madhouse in daytime, as quiet as a cemetery at night, it included the shipping center. All the shipping companies were located in or around Broad Street, off Wall Street.

I was staying at the Lutheran Seaman's Club. Mrs. ... (can't remember her name, but see her face clearly) was a sort of denmother to us seamen of all nationalities. The Lutheran place was a lot quieter than the big Seamen's Center down the road a little way, with its nickname of "The Doghouse." Drunkards and winos flocked round the Seamen's Center, so it was worth paying a little more for a bed and room at the Lutheran place.

Mrs. ... (gotta recall her name -- I'll dismiss the problem and hope the old subconscious throws it up suddenly) was very sympathetic. When, in my futile ship-to-Japan search, I started running out of money, she got me in contact with employers who weren't fussy about U.S. immigration laws. That kept me going, but there was no real hope.

Then it happened....

Late one afternoon I was arriving at the club, and Mrs. Doo-dah (oh, God bless me, I *must* remember her name) was standing in the foyer -- smiling, smiling -- and waving a paper at me. It was a telegram from a sister-club for seamen in San Pedro, California, with whom she'd been in contact:

> NEED URGENTLY RADIO OPERATOR CHIEF
> ENGINEER AND FIRST ASSISTANT ENGINEER
> FOR LIBERIAN FREIGHTER BOUND
> ALASKA AND JAPAN

When I'd read this and looked up I saw Mrs. Whatsername smiling through a golden haze -- and two smiling faces beside hers -- two of my fellow boarders at the Lutheran -- both German, both engineers and obviously the happy folk who fitted (with me) into the needs of the telegram.

When we'd done the obvious immediacies like phoning the airport to reserve seats for three to Los Angeles and started putting our things into bags, I wandered outside for a stroll.

As mentioned, this part of Manhattan was (and is) either in a frenzy or else dead and deserted. There are, however, two in-between periods every morning and night, when most people have gone home but there are a few stragglers catching up. To me there was a golden haze on everything. I walked with a bouncing spring in my step up Broad Street, smiling at everyone I met. How jolly was the music coming out of that jukebox in a luncheonette! How pretty that smart secretary was! ... and that one ... and that one! A cop standing at the corner of Broad and Wall grinned at me. I grinned right back -- then I had real cause for laughter when I saw the statue of George Washington outside the Federal Reserve Building; he was holding his hand out in a sort of peace-making gesture but it was also the exact stance he'd take if he was bouncing a yo-yo up and down. Someone had actually *put* a yo-yo on the wrist of George's outstretched arm!

The music, the pretty secretaries, the jovial cop, George Washington and his yo-yo -- they all were in league with me against solemnity, sadness and the serious business of moneymaking. I had my prospect of Japan and Sumiko. I, they, we, everybody was happyhappyhappy.

* * * * *

As we flew to L.A. I was busily writing to Sumiko and trying to calculate very approximately when I could hope to be in Japan. When we eventually got to the ship, the *James Lick*, we discovered that her middle name could have been *"Trouble."* She had been under the American flag, then sold and transferred to Liberian registry. We had to go through a line of pickets (set by U.S. unions) to board the vessel.

There was I, with a score of other foreigners, in immediate need of a job. Two years ahead I'd be a U.S. citizen, chauvinistically resentful of foreign sailors taking U.S. jobs! Which viewpoint was right? It's all relative -- just as Einstein said. Well, the captain of the *James Lick* found an easy way out of confrontation with irate ex-crewmembers and union agitators. We raised anchor in the middle of the night and slithered away -- northbound for Alaska (thence to Japan).

That wasn't the end of the *James Lick*'s labor troubles. She had made headlines on an inside page when I joined her; she'd get front-page headlines when I returned in her to Portland, Oregon, a thousand miles up the coast. For now, though, all that mattered to me was that the ship was seaworthy and would be heading, sooner or later, for Japan.

We were a mixed bunch. Captain and deck officers were all American. I was British. The top engineers were my New York comrades from Germany. The rest were Spanish, Greek, Puerto Rican and one Chinese. The Old Man seemed to be a good-natured villain. I didn't give a damn about anyone or anything. I was going to hold Sumiko in my arms again and savor the civilized atmosphere of Japan.

We went to Seward, Alaska, to load stuff for Yokohama. It was bitterly cold. I went out for a walk up a nearby hill on the edge of town. As I was coming back, I fell over a small mound and tumbled into a hollow. It was a fall of only a few feet, but I landed heavily on my side, with my arm twisted under me. I felt considerable pain, but I pulled myself together, straightened out and up. The odd way my left wrist was dangling made it look as if I'd busted it. The sharp, stabbing pain in my chest suggested that I'd broken a rib, too.

Physical discomfort was nothing compared to the horror and frustration that swamped my mind. Had I gone through all those months of search/inquiry/disappointment/misery/despair, then been sent soaring heavenward by the sight of that telegram, then experiencing every person, thing and incident since then as individual instruments in a huge orchestra in the final movement of, say, Beethoven's Fifth Symphony--all leading triumphantly to the reunion with Sumiko, only to have it all be a house of cards, brought tumbling down by one stupid misstep on a hillside?

The shock, both mental and physical, had to be dealt with immediately. Seward is a little town. I soon found a bar, and croaked my desire for a double brandy. The barman quickly got me one and had a willing ear for my story.

"Look, feller," he advised, "thing to do is to claim you had the accident *on the ship.* Let the company pay for your treatment. If you get stuck with the hospital bill they'll take all you've got!"

The cost of anything at that time in Alaska (shortly before her statehood) was astronomical. The brandy I was drinking cost three times the stateside price. I appreciated the barman's advice. He might have been wrong, of course; the shipping company may have been responsible for any medical attention I needed, no matter where and how, but best to play safe. It was lunchtime when I went back aboard the ship. I entered the mess hall where the captain and officers were starting to eat, pointed to my peculiarly dangling wrist, and said I'd just slipped on the icy gangway. I was taken to the local hospital. They took x-rays, and wanted me to stay overnight in the hospital for evaluation.

I had misery seeping through every pore. "Am I to be taken off the ship?" "Can I be patched up and go back on board?" were my repeated questions to doctors and nurses. The answer was always the same, "Sorry, we don't know yet."

My wrist was put in a cast and my left arm in a sling. They put another over my cracked (not broken) ribs, but the doctor told me, "Look, you're going to have a stab of pain every time you breathe, no matter what. This plaster cast over your chest isn't really doing anything. It constricts your breathing and you'll be begging me to take it off. That's okay by me, and...."

"Okay, doctor, but all I want to know is...."

"Yes, yes. All you want to know is -- can you rejoin the ship. Wait! We'll know tomorrow morning."

It was as he said. The cast on my chest gave me claustrophobia -- I wanted it off. Then he told me that my wrist cast would have to be taken off by *Japanese* doctors -- which was his way of telling me that there was nothing to prevent me rejoining the ship! My anxiety drained away like water noisily draining out of a bathtub. Sumiko, here I come! I was lucky, lucky, lucky. If it had been my *right* wrist, I couldn't have operated a Morse key!

Thus *I* got relief. Not so the Old Man when he saw the bill the hospital presented. Seward was "up against it" financially. The town used to save money by rationing even the electric power -- turning it off for hours. They weren't going to lose an opportunity in my case -- x-rays, overnight stay in the hospital, casts, etc., came to $874!

Joyfully I wrote to Sumiko, now able to make jokes about what had driven me nearly crazy. I told her that as soon as we got to Yokohama (the port for Tokyo) I'd send her money to come up to the Big City by train -- and I wanted to see her dressed old-style -- in kimono, etc. We were going to be some weeks there -- no rush, no fuss, no tanker tempo.

I filled in the short daylight hours in Alaska (we weren't far from the North Pole, and it was winter) by this and that -- any old pastime, while my injuries healed. The stab-stab-stab (every breath) in my ribs grew less and less. My left forearm, with its massive plate of plaster, was carefully shielded.

At long last, anchor went up, ship's horn hooted, and we "set sail" (remnant of sailing days). It was still a long, long haul across the Pacific, in a ship that only did eight knots, but for once the "peaceful" ocean lived up to its name. (Balboa's christening of it one lovely summer's day in 1513 was just about the greatest misnomer ever, when the terrible typhoons -- the worst on earth --are considered.)

We docked in the hustle-bustle of Yokohama, and I was soon engineering Sumiko's first visit to one of the most densely populated cities on earth. The getting-rid-of-my-arm-cast by a Japanese doctor was a minor detail in the many, many ups-and-downs since I'd left Sumiko -- how long was it, now? -- eight months ago!

Sumiko's train would be arriving at 11:07 AM. I was there with what should have been time to spare, but the language barrier and the beautiful but incomprehensible pictograph signs had only put me in the direction of the right platform by 10:55 AM. The huge railway station was almost deserted (rush hour was past). In my imagination there should have been soft background music as Sumiko and I saw each other from a distance along a nearly empty platform -- and we would run toward each other, to collide in an ecstatic embrace.

It wasn't quite like that. I saw her first. She had the vast superstructure of the station as a frame for her as she stood, alone, at the top of a ramp. She had laid down the bag she was carrying, whether to rest or to look for me, or both. She was the country girl, come up to the big city to meet her lover. Her appearance, as she stood at the top of the long ramp, was graven on my mind forever.

My eyes traveled down from her head to her feet. Her hair was piled high, with a red brooch at the top. Her pretty little face had a touch of makeup. She had dressed herself from head to feet in the formal, old-style manner I had requested, and was very picturesque. Part of that setup was that her bosom was tightly constricted by the huge red sash wrapped around her -- but that was okay for appearances, now. Her delicious little breasts would be exposed and plundered later. Her colorful kimono met the ground at her feet and completed a delightful picture. As in the Japanese paintings, where a few brushstrokes on a nearly empty canvas emphasize the appeal of what is shown, so her tiny figure gained from the huge surrounding station. In a word, she was irresistible.

We collided in the expected blissful embrace. "Oh," she whispered, "it has been so long. I no see you for year."

It wasn't that long, but emotion and the mind, as we all well know, s-t-r-e-t-c-h-e-s time.

"I not go," she said, "with anyone else -- nobody."

I took her back to the ship. The captain, most of the officers, many of the crew -- nearly everyone had his "wife" for this port. This was really going to be a happy ship. The softening effect of women was going to round off all the edges, lessen impacts, sweeten routine miseries, and generally make life worthwhile.

Sumiko and I had a ball. At that time one could have a good time in Tokyo for little money. Each day, after we'd finished touristing, eating, dancing, going to shows, etc., we'd return to the ship to make love -- after putting a DO NOT DISTURB sign on my cabin door. We might not get up until nearly lunchtime.

I never spent any money on hotels with Sumiko. If we weren't in her own village -- with her friends or relatives, the ship was our hotel.

You may have noticed that the word "love" has not occurred anywhere until a few sentences ago, and the expression "making love" is an euphemism that fools nobody now -- not that genuine heartfelt love is excluded from it.

Let's return to that moment when she stood at the top of the ramp in Yokohama railway station. It was The Climax. Everything had led up to that moment; everything thereafter was a falling-away (at first gradual, then steep). It was the ultimate reward for all my wheeling and dealing, my patience, and my impatience. After which self-praise, I retreat in shame from the way I was *using* the poor woman. My chauvinistic machinations -- were they at her expense? They didn't strike me that way at the time, but I certainly wasn't sensitive enough to her point of view. There must have been considerable excitement in her village about this Westerner who was so smitten with Sumiko that he was moving heaven and earth to rejoin her. She was going to be rescued, Cinderella fashion, and would start life anew in Europe or America.

What was the brutal truth, from *my* angle? I was a sex-starved sailor who had a bit of good luck in a remote Japanese port one night, meeting a young female also single-minded. We had a good time together. When I had to go, I resolved to come back. After considerable trouble, I did. I didn't have any real urge to go on from there, and anyway the tide of affairs/circumstances/other people/jobs swept me well away from the idea of making my sweetheart of one port into a wife for always.

I was very fond of good-tempered, fun-loving pretty little Sumiko but I'm glad we didn't have a tidal wave of mutual passion that would have encountered mountainous obstacles. At that time I was almost without a country; I was still a British citizen, though not living in the U.K. East is East, West is West -- and, though marriages often do unite them -- oh, enough said! The dismal prospect of again having to make all those ponderous efforts to get one ship making one voyage -- brrrrr.

Shortly before the ship was going to haul anchor for a return to the States, I took Sumiko back to the railway station. She recognized some folk from her town who were also going home -- and feeling suddenly formal and proper, she wouldn't allow any signs of affection -- certainly not kissing. We had vague plans for the future, but they were as insubstantial as the mist around the summit of Mount Fuji.

When the *James Lick* reached Portland, Oregon, it didn't take long for the ship to get onto the front page of the local newspaper. She was in the process of being sold again to another company: officers and crew had beefs about pay. The engineers (led by my two cohorts from New York) were threatening to go on strike and not keep the boilers going -- in freezing weather. There were strong hints of corruption, finagling, bribery -- it was a mess. The U.S. Customs & Immigration (even they were not immune to charges of corruption) cut the Gordian knot and flew all us foreign seamen as far away as they could -- to New York, for deportation. We were being held in a detention center in downtown New York (not very far from the Lutheran Seaman's Center).

I got hold of a typewriter and banged out a plea for us foreign seamen, attacking (best defense) all the companies, authorities (even the U.S. Customs & Immigration) on the West Coast who'd been giving us a hard time. It convinced the New York authorities. They released us. I went to a travelogue movie in Greenwich Village and there met the woman who'd eventually become my wife. Another story begins. Fade-out, fade-in.

I don't want to enlarge or diminish this crossing of my path with Sumiko. Here it is, because this is the way it happened.

For me, always, she'll be standing at the head of the ramp in Yokohama railway station, dressed in a colorful kimono, her bag by her side -- looking everywhere for me.

RANGES OF { EYESIGHT / EARSHOT

"ON A CLEAR DAY, YOU CAN SEE..."

Up on the flying bridge (or "Monkey Island") you don't see "forever," but the feeling of spaciousness on a clear day is grand. The wind beats your face, and you look across to the horizon, every which way. No ships. Only the tops of waves breaking -- little white spots of foam -- here, there, all over.

Some mathematical calculations help fix all this. The height of the flying bridge being 53 feet above sea level, the view to the horizon -- the radius of the circle -- is 9.2 miles. That's all. The circumference -- all the way around the horizon -- is πr^2 -- 270 miles. You can see about 9 miles out in each direction if visibility is perfect. When there's rain, mist or fog, visibility can descend from 9 miles to 9 yards.

"ANY DAY, YOU CAN SEE..."

Down the deck to the wheelhouse is only a few feet. The visibility range isn't much less. Here the eyes of the navigating officers are used strictly for business, navigation and safety, not sightseeing. An engine room telegraph (FULL AHEAD -- HALF AHEAD -- SLOW -- STOP) tells them down below what the navigators want. There's a radar here. The visual range of that is 40 miles, any direction, any weather, even dense fog. Little light spots on its round screen show anything solid: ships or land masses. This magic box -- a recent innovation on ships (only 30 years or so) is an electronic development. The main electronic center of the ship is at the back of the wheelhouse -- "Certified Radio Room."

"ANY DAY, YOU CAN HEAR..."

The radio console looks impressive, with all those knobs and switches. Looks are unimportant, though. Sight doesn't count but sound has an almost unlimited range. Through those earphones or that loudspeaker come sounds from the other side of the world. That receiver knob -- put it in a certain spot, and Hong Kong comes booming in, maybe with a message or requesting same from you, here. One-tenth of an inch from that receiver-knob position, the receiver is tuned to somewhere 8,000 miles from Hong Kong -- WCC/Chatham, Massachusetts. That place, too, ready to communicate with this radio shack. Push the knob one-sixteenth of an inch the other way--we have leaped across 3,000 miles of the Atlantic Ocean, and Portishead Radio, near Southampton, England, is calling, requesting GKU contact.

The range, in the radio shack, could be 10,000 miles.

"I'll put a girdle round about
the earth in forty minutes."

-- Puck

A Midsummer Night's Dream

"Heck, I can do
that in .053681
of one second."

--Sparks

SS Anyship

WENDENS AMBO Essex, England

Ironically, I'd gone into the bar to escape the heat. (My ship was in New York, and I was wandering around.) You know about New York City's perennial coiled-spring tension? It's the attitude of an irritable cabby frustrated by a red light, squeezed in every which way by skyscrapers and surrounding traffic, but still rarin' to go. You may not know about the heat in August. It's hell. The banner outside the bar promised "AIR-CONDITIONING INSIDE!" So in I went.

There were so many people inside that the Fire Commissioner's red sign on the wall (OCCUPANCY LIMITED TO 145) was just a joke. The yelling, smoking, sweating mob was giving the air-conditioner such a hard time that the only relief was going to be a glass of cold beer. One of these arrived after a dangerous journey from the bartender, over shoulders and around backs. I took a long swig and paused. Then I saw, about 10 feet away, the calendar picture on the wall.

It was a color photo ... country scene ... European, probably English: four thatched cottages several hundred years old and a church hiding behind a tree at the back. Splintered sunlight came through overhanging tree branches. The atmosphere was that of gentle warmth, not the oppressive humidity of New York. Two people -- obviously tourists -- were strolling by the cottages. Probably the implication was that they'd been brought there by Pan-Am Airways (whose calendar it was). I certainly wished I was there with them, right then.

I wanted a closer view. I "excuse me'd" and pushed and wriggled until I could see more details. The white-walled cottages were dappled with sunlight and the shade of leaves. The squat, little church -- gosh, it looked very old -- brooded in the background. The tourists fitted unobtrusively into the scene, and altogether it was quite a skillful picture. Taken where? The corner told me: WENDENS AMBO, ESSEX, ENGLAND.

I smiled. "Ambo" (whatever it meant) was quaint. "Wendens" (whoever/whatever he/it was) had pushed me to the point of no return. I just had to get a copy of that picture!

I didn't expect it to be easy, but opportunity was literally just around the corner. At that time, the new Pan-Am building was being constructed in midtown, a few blocks away. It was mostly scaffolding, cranes and men with hard hats, but some offices were nearly finished on the 52nd floor. Soon I was in a high-speed elevator heading there. I whizzed by girders, planks, bricks, etc., impressed by what I'd heard: that this was to be the largest office building in the world. (At that time, the World Trade Center was still being planned.)

The Pan-Am folks on the 52nd floor received me courteously. Yes, certainly I could have one of their calendars, if and when they could locate one. Would I please forgive the mess? I waited among still-unopened packing cases.

I was peering down at the tiny, yellow ants which were cabs on 42nd Street, when a smiling girl, a pretty little redhead, returned with the calendar. Thanks, small talk, polite exchanges, and I left. Back to the elevator. Even as I was swallowing to relieve the pressure on my eardrums, I was separating Wendens Ambo from its monthly companions. Pan-Am had provided a portfolio of scenes, one for each month: a panoramic vista from the statue of Christ overlooking Rio de Janeiro, all the spectacular places to which Pan-Am could take me. Sorry, not interested ... ground floor at last ... trash can handy -- away with the Eiffel Tower, the Grand Canyon and all the other spectaculars. I had what I wanted.

Back on my ship, I installed Wendens Ambo on the portside bulkhead of the radio shack. On the starboard side was a porthole, so two pictures faced each other: one was circular and real life; the other was this rectangular sample of a summer's day in rural England.

A year passed. The view through the porthole was usually the sea, of course, but the colorful port scenes added variety. In Karachi, coolies with sacks scurried around huge hills of grain which we'd just brought and unloaded. In Rastanura, there were billowing clouds of sand partially obscuring the pipeline snaking out of the desert. Another time, clouds were whirling snow and sleet, whipped up in bitterly cold Gdynia, Poland. None of these circular pictures affected the rectangular one: the tourists were forever passing the cottages.

A vague resolution germinated all year. Then one day in Djakarta (East Indies, near the equator), when the temperature was 110 degrees in the shade, I looked up at those never-changing cottages, and my vague wishes flowered into a decision. Dammit, I was going to go there!

I owed that part of the world a visit, anyway. My parents and my sister with her family lived in Essex. It took another year of fruitless finagling, trying to get a ship going to England, before I gave up. I decided to take a vacation and go as a *bona fide* tourist -- and as Pan-Am passenger, of course! Poetic justice required it.

So the long-studied calendar picture became a three-dimensional reality as I hove onto the scene at long last in my brother-in-law's car, surrounded by his family. We had a camera ready. I posed alongside my nephew and niece, with the cottages as a background. Snap! They posed with my sister. Snap! I stood alone in front of the church. Snap! Now I had evidence: Kilroy-was-here stuff.

The photos came out well. I pinned them to the original calendar picture on the bulkhead and they went globe-trotting with me as the round picture (the scene through the porthole opposite) changed and changed. Wendens Ambo had changed for me, too, though not in appearance. No longer a remote place, it was part of a personal experience, and memories of it merged with the impact of the picture on my eyes. It was part of me. Whenever I visited England, I'd always pop over to Wendens Ambo. Family kidded me about my pilgrimages to the spot.

Sometimes, as I looked at the picture, I wondered about other lives that place had impinged on, merged with, affected, during 400 years. The dividing lines between years, even centuries, seemed to blur.

... It's 1851. Up the lane to Cottage Number One comes an ex-merchant of the East India Company. He left the stagecoach half a mile away, and has come to claim this place for his retirement.

... Now it's 1765. Knocking at the window of Cottage Number Two is a sailor, son of the widow within. He's got plenty of souvenirs in his duffel bag, and he'll tell her about Boston, in New England, where they're raising hell about the Stamp Act.

... It's 1645. There are discreet knockings at the door of Number Three, then hoarse whisperings at the window. Inside is a lovely woman. Outside, her cavalier husband is on the run, his horse sweating and panting by a tree. The king's men have been badly beaten at the Battle of Marston Moor. He's off with others to accompany Prince Charlie on the ship to Holland.

... It's 1588. The pleasantly plump farmer's wife at Number Four has little interest in those rumors about a Spanish Armada. A hundred ships or more (they say) have been wrecked all around the coasts of England, Scotland and Ireland. She is satisfied by the housecleaning she just finished. She's going to sit outside and chat to her neighbors until her hubby comes home from the fields for supper. The cottages behind her look pretty much the same as they will be on a picture in the radio shack of an American freighter as it bucks the Humboldt Current in the Pacific Ocean 391 years later.

The Elizabethan was smug and snug in her here-and-now. The nervous cavalier had his here-and-now. The cheerful sailor had his focus of attention, the things that made sense to him. The thoughtful merchant had his. I had mine. Centuries blur. Here-and-now becomes there-and-then. Steady factors through the years are the church and cottages.

Enough of them. No more fantasy. Might-have-beens are evanescent. Let's turn to that quiet background and actually visit the church. There will be records -- dates, facts, people, sequential history. Maybe the vicar will know a thing or two and be glad to help....

Yes, indeed. He and the records soon put us straight.

"Wendens" means "beautiful valley" or "winding valley." "Ambo" is Latin for "both." There were two parishes, Great Wendens and Little Wendens. "Both Wendens" were united in the present church in 1662. (Little Wendens church has disappeared.) As a further distinction, the old separate villages are spelled "Wendons," while the modern one is spelled "Wendens."

The church is twice the age of the cottages. The first church on the site in the eighth century was probably of wood. It was rebuilt in stone about 1086 A.D., soon after the Normans conquered England. The tower is as it was then, almost unaltered. There were other alterations -- arches here, windows there -- every hundred years or so. The beautiful organ is comparatively modern -- dating from 1780 A.D.

The registers commenced in 1540 A.D., just about the time the first cottage was being built outside the west tower. Even earlier is a stone slab in the southern wall, bearing the brass of a man in plate armor. This commemorates the patron of the church in 1410 A.D. Memorials and slabs to other people through three centuries are outshone by a tombstone near the south door in memory of William Nicholson. He was a midshipman in Nelson's *Vanguard* (this was prior to the Battle of Trafalgar in 1805). Nicholson lived on until his very ripe 104th year, in 1886.

* * * * *

We've dashed around plenty through time and space since I first saw the calendar picture through smoke and haze in a New York bar.

What about 100 years from now? Science fiction already has a tough job keeping abreast of achievements. By that time (with trips to Mars routine), a rocket might be drilling its way into outer space, carrying passengers in a state of suspended animation. After 20 years of traveling thus, on tachyons, going faster than the speed of light, they would be revived by remote control, to find their bodies had aged only three months.

Meanwhile, back on earth, or a darned sight nearer to it and everyday life, Pan-Am may still be shuttling people from country to country. If they reprint that calendar late in the 21st century, perhaps some enterprising mariner of the stars will tape it to his bulkhead, next to the diamond-hard porthole that looks out on Andromeda, Orion's Belt and the rings of Saturn.

Down in Wendens Ambo it'll still be pretty much *is* as *was*. A hundred years is trivial to the church. Its solid foundations and stonework should last another thousand years. The western wall of Cottage Number One will probably have subsided an inch or two. It might need buttressing ... but not much.

* * * * *

In Saffron Walden I delightedly found a just-published book, *Wendens Ambo: The history of an English Village,* by John J. Mackay, a Scottish insurance broker in the City of London who had come to live in W.A. He had fallen in love with it, as I did, at just about the same time. We became instant friends. His book unrolled 1,000 years of history and brought to life vicars, farmers, villagers who had been interesting and significant.

One day we rounded off a tour of W.A. in the loud, cheerful atmosphere of the Bell Inn, where the camaraderie of the villagers echoes that of their forebears 900 years ago when the church and the pub where both being built. The Saxon bricklayers (supervised by Norman architects) slaked their thirst with mead, beer and wine. Their jokes and glass-clinking mingle with those of many, many others down through the centuries. Let all merge into a single, here-and-now-, noisy, smoky stream going up through the magnificent 300-year-old chimney. Let this stream join sounds coming through "bell-holes" on the steeple, hover round the famous church, and look down on W.A. houses. Most of them look as they did long before 1700 AD. Some date to before 1600 AD.

1500 feet up is the favorite spot for aircraft from *National Geographic* magazine to get an overall view of places like W.A. We see all the village, from merrymaking at the Fighting Cocks (the pub in the north) across and down to similar high jinks in the Bell just below. Over there, in the moonlit churchyard, is the lichen-covered tombstone of William Nicholson. Ghosts of other notables swirl upward to us: John Feltwell, a troublemaker of 1591 – he was seditious but he luckily got off with only two hours in the pillory ... Charles Barnes (vicar, 1892-1905) renovated the church with much TLC and died aged only 43 ... All sorts of tales and people ... Helen Keller, the most famous deaf-blind person ever, had a rest-cure here in 1930. Her loving memoirs back up a universal opinion: Wendens Ambo is an unforgettable, lovely little village.

Hero/Villain GEORGE ROGERS and the MORRO CASTLE 1934

One of history's worst maritime disasters commanded world attention in September, 1934. The *Morro Castle* became a flaming torch 20 miles out from New York; 135 of its 549 passengers and crew died. Out of the smoke and flames at that time, and through the haze of unanswered questions ever since, there emerges the gross, blubbery, pear-shaped figure of the chief radio operator, George Rogers. He was an undisputed hero. If he hadn't stuck to his post as he did (though being overcome by fumes and heat), several hundred more lives could have been lost. He was feted by parades, dinners, awards, medals, etc., and even went on a Vaudeville tour for five times his pay as a radio officer.

Yet circumstantial evidence, plus logical suspicions based on his record before and after the *Morro Castle* fire, indicate that he was also the villain who started it! Solid evidence? Eyewitnesses? Confession? Proof of any kind? No, all are lacking. Nevertheless, Rogers had been lying, stealing and connected with mysterious fires from the age of 12. Later he twice served prison sentences for attempted murder. His criminal record encompassed 44 years, yet there was never anything more than circumstantial evidence against him! Toward the end of his life, in 1956, he was bluntly asked: "Did you start the *Morro Castle* fire?" His answer was evasive.

Even Agatha Christie wouldn't have dared assemble so many clues to a whodunit without final proof and certainty of guilt. In 1972, Gordon Thomas and Max Morgan Witts wrote Shipwreck; the Strange Fate of the *Morro Castle. Theirs is far and away the best book on the subject, but even they are a little remiss in some technical details about the sending of an SOS -- details which only a radio officer would spot. This account will fill in a few gaps, and try to combine a bird's-eye view with a radio insider's viewpoint.*

George Rogers is an intensely embarrassing subject for any fellow radioman. He really did earn the Veteran Wireless Operators Medal awarded him, but he also really was the arsonist who killed 135 people and destroyed a ship. He was the hero and the villain!

* * * * *

In 1934, the Depression overshadowed just about everything. Nevertheless, Ward Line could still find people to go on a "whoopee cruise" from New York to Havana and back. The *Morro Castle* and her sister ship, the *Oriente*, were affectionately called the "Havana ferry-boats."

The *Morro Castle* was a floating gin mill, a playground for the rich, an escape hatch. On this, her 174th voyage, Captain Robert Wilmott (her skipper since her maiden voyage in 1930) had reason to suspect someone was trying to poison him.

That was only one of his troubles. There had already been a fire aboard, an attempted strike and suspected sabotage from Communists in Havana. Wilmott linked them all together, shunned socializing, kept to his cabin and was regularly seen by only three people. These were the chief mate and the chief engineer (with news from their departments) and the chief radio officer, George Rogers, bringing telegrams from the radio shack. Circumstantial evidence points to the opportunities Rogers had to poison the captain.

Whoa! This is supposed to be only the background. The people involved don't rightly belong here. So we'll drop the curtain on their activities for a while and concentrate on the background conditions, the ship itself.

Traveling first-class on the *Morro Castle*, the passengers could celebrate like Elizabethan England after defeating the Spanish Armada. In the dining room were flutes and mandolins on the bulkheads. The first-class smoking room resembled a drawing room in Versailles when Louis XVI was king. The trip cost from $80 (third-class) to $160 (for a first-class cabin). It was like putting to sea in the Waldorf-Astoria. Yet it was as false an exterior as the biblical "whitewashed sepulchre": the *Morro Castle* was a floating fire hazard.

Lack of safety precautions was the theme of many a discussion in the crew's mess. There were no real fire-and-boat drills; in fact, the fire-detecting system didn't even work. Both omissions can be blamed on Captain Wilmott. He, in turn, could blame the company policy of DON'T DISTURB/OFFEND THE PASSENGERS!

On an earlier cruise, a leaking hose connection caused a woman to fall and fracture her ankle. Ward Line settled out of court for $25,000. Captain Wilmott, to stop further trouble, had the deck fire hydrants capped and sealed. The fire hose was locked away, along with nozzles, wrenches, etc. The ship became nothing less than a fire hazard from stem to stern. The only fire equipment available were some portable fire extinguishers. The required fire inspection of the ship had taken place a month before the alterations so there was no chance of this extraordinary situation being officially detected.

Fire drills were only charades. A 42-foot length of hose was just dragged noisily along the deck. There was no way to test it -- no water was available through the fire hydrants.

The fire-detecting system was a sophisticated assembly of pipelines leading from the cargo spaces to a detector cabinet in the wheelhouse. An exhaust fan sucked up air and immediately spotted any smoke below. It was a beautiful, almost foolproof device. Unfortunately, a cargo of salted hides had been loaded, and the stench was awful. Captain Wilmott remedied this situation by simply turning the detecting system off! Possible complaints from passengers with sensitive noses were scotched.

There were other dangers aboard: fire doors without sirens or bells; the gangway watchman had no idea where the nearest fire alarm was; some lifeboats were almost useless because no one could get into them from the promenade deck. Worst of all, the Lyle gun (a line-throwing apparatus) had been taken from the bridge, where it belonged. Captain Wilmott decided to move it "as it might get some excitable Cuban into thinking we were an armed ship." Where to move it? Third Mate Hackney found an empty compartment between the promenade deck and the ceiling of the first-class writing room! The Lyle gun and the *drum of gunpowder to operate it* were placed there. They were easily accessible from the bridge in an emergency but there was only thin plasterboard between 25 pounds of dangerous explosives and the writing room.

Captain Wilmott didn't know that he was simply paving the way for a dedicated arsonist. Time now to switch from the ship to the people.

MASTER: CAPTAIN ROBERT WILMOTT. The skipper is the logical starting point in any nautical story. He is the personification of the company and/or ship. He's the dictator (even if an amiable one) at sea. We've already had a sneak preview of Captain Robert Wilmott's troubles. Born in England, he'd started seagoing as deck boy in 1902. By 1930 he'd been given command of Ward Line's flagship (just launched) the *Morro Castle*. Just recently, in 1934, he'd married a first-class passenger. She was a widow who (thanks to the regular schedule of the *Morro Castle*) could spend every second weekend with him. He had all the outer signs of glamorous success and happiness, but, as noted, he'd become depressed -- a recluse in his cabin -- drinking only bottled water.

He had unburdened some of his fears on Captain Oscar Hernandez, Havana's chief of police, who agreed that his life and ship were in danger. The previous fire, possible poisoning and strike threats were all "classic symptoms of Communism," Hernandez had said.

Wilmott's suspicions then focused on two men in the radio shack. His uneasiness about George Rogers was instinctive, not based on solid evidence. "That's a bad, bad man," he'd whispered to cruise director Robert Smith one day as Rogers was approaching. He had told one passenger he was going to fire Rogers in New York. More particularly, though, hard evidence had crystallized around George Ignatius Alagna, the first assistant radio officer. After the Havana police chief had talked to him, Wilmott called Chief Mate William Warms. "Bill," he said, "there's a Red in the radio shack. Get some irons." Warms had argued against this, even though he, too, disliked Alagna. (Bad publicity, bad effect on passengers, etc.) So Wilmott agreed just to keep an eye on him until they got to New York.

CHIEF MATE WILLIAM WARMS. He'd been at sea since the age of 12, when his salary was two meals a day and a dollar a month. He was always a hard worker. He climbed the promotion ladder and in 1918 had become the captain of a Ward Line fruit boat. There seemed to be no reason why he should -- as he did -- ignore Rule One: guard the *safety* of passengers (if any) and crew. As captain of the fruit boat, he'd never had any fire and boat drills.

Well, some crew members complained to the right people; Warms lost his license for 10 days, and Ward Lines beached him for a year.

Then he'd been given command of a Ward cruise liner. Two mysterious fires broke out. Again, fire precautions were found lacking; again Warms lost his command; again he was demoted and beached for a year. He swore to his dying day that he'd been made a scapegoat.

Anyway, he'd been disgraced twice, and it was now up to him to compensate for past deficiencies. He did just that. He established a reputation as the best cargo officer in the U.S. Merchant Marine. It was to be supremely ironic that lackadaisical or nonexistent fire precautions on the *Morro Castle* -- not directly his fault -- should be partly responsible for one of the most spectacular ship fires ever. He was her captain for one of the shortest commands ever, 15 minutes under 24 hours.

CHIEF ENGINEER EBAN ABBOTT. He and Chief Mate Warms had been "daggers drawn" enemies since the maiden voyage four years before. "That worm on the bridge," as Abbott called him, returned intense distaste for "that stuffed tailor's dummy in the engine room," according to Warms. Captain Wilmott had been observing this increasing tension and had just decided to do something about it. He told Warms he was going to recommend to Ward Line that Abbott be transferred to the sister-ship *Oriente*. Enough of Abbott for now. When the ship is in real trouble, his will be a shameful story.

GEORGE ALAGNA, FIRST ASSISTANT RADIO OPERATOR. The man who Captain Wilmott considered a possible Communist troublemaker actually had no Red sympathies whatever. George Alagna had already worked on board the *Morro Castle* for four months, on a lease arrangement from RCA. Those four months had made him extremely dissatisfied with conditions under Captain Wilmott. He prepared a list of grievances, headed by a description of the food as "pig swill." Rogers had seemed sympathetic to the protests in conversations between the two, and had urged Alagna to organized a strike just before the ship was due to sail from New York. (The ship couldn't sail without a full complement of radio officers.) Rogers would be unable to participate in the strike himself, he said, because he was involved in "undercover work for RCA."

Alagna canvassed the junior officers and other crew members but found no support for a strike. In fact, the others began to shun him for his near mutiny.

Captain Wilmott heard the news and wanted to fire Alagna on the spot, but he couldn't get another radio officer on such short notice. The ship sailed with Alagna still second in command in the radio shack, but under suspicion by nearly every crew member on board.

The situation grated on Alagna's nerves. He began to think someone was trying to waylay him; he imagined footsteps behind him on the dark deck. Once he raced up to the bridge and accused the watch officer of tinkering with the radio-compass and thus jamming the main radio transmitter. It was a ridiculous, impossible idea, but it showed Alagna's nerves were wobbly, inflammable. This incident sealed his fate. Captain Wilmott sent a wire to RCA from Cuba, demanding Alagna's removal when the ship berthed in New York. This was going to be his last voyage on the *Morro Castle.*

SECOND ASSISTANT RADIO OPERATOR CHARLES MAKI. He was a 19-year-old Finn whose main concern was body-building. Pictures of men with bulging muscles surrounded his bunk. His importance in the *Morro Castle* story is almost zero. He is a contrast with the most important figure --

CHIEF RADIO OPERATOR GEORGE WHITE ROGERS: He was born in 1901 with a rare disease, a pituitary disorder called adiposogenital dystrophy, which surfaced at the age of seven. His whole abdominal region began to swell up. His hips, thighs, upper arms and back put on 50 pounds inside a month. The thinness of his lower arms and legs only accentuated the disproportion. He was raised in Oakland, California, by his maternal grandmother, since both parents died when he was five years old. Poor George -- a pitiful sight in cut-downs of his father's clothes -- was taunted by neighborhood kids. He had knock-knees, flaring hips. He was an object of ridicule. At the age of 12 he weighed 170 pounds.

A classic symptom of the illness is sexual underdevelopment. George had feminine, hairless skin, and his voice never broke into a manly register. He was fat. He was secretive. As a young teenager he began to steal things.

He started with a pair of skates (pointless thievery -- he was too clumsy and off-balance to use them.) He was let off with a warning. He progressed to burglary and was sent to reform school. There he committed petty thefts and received a long series of bad reports: "untruthful," "bad influence on others," "a moral pervert."

The Boys and Girls Aid Society in San Francisco tried firm discipline in 1915, when he was 14. It didn't help. He just became alternately silent and then domineering.

In this downhill (and momentum-gathering) progress of George Rogers's condition and behavior something emerges which was *not* his fault; it's rather an accusation against the authorities who had been watching his disintegration. Not once was he seen by a doctor or a psychiatrist. Either would have recognized the symptoms. Both together might have done him some good.

Almost by accident, George took to radio. Here was hope. The San Francisco society found him a job as an assistant wireless operator. He was paroled. On May 12, 1917, he went to sea on a schooner sailing out of San Francisco. It looked as though things were really on the up-and-up for George at last. He'd found a job he liked; he was good at it. There was a niche in life where he could function as a useful member of society.

Had this disturbed personality experienced a profound change for the good? Doubtful. On the contrary, after 12 years of seagoing he found a new and absorbing interest -- arson. One special attraction of this crime -- to those with twisted minds and steady resolve -- is that, if done properly, it totally destroys all the evidence.

It was 1929. He had been experimenting with fountain pens filled with acid which ate through membranes of copper. He was employed by the Wireless Egert Company in New York. A mysterious fire broke out early one morning. Rogers had, for some unexplained reason, come to work an hour early, and he obligingly let the firemen in. He was under suspicion (as he would be frequently in the future) but nothing could be proven. "Mysterious fire" and "circumstantial evidence" are unwieldy polysyllables, but they're going to occur so often, we'd better shorten them to "m.f." and "c.e."

A set smile (his trademark) was frozen on his face as the police grilled him, off and on, for days. Salvage experts felt sure that a chemical timing device had started the blaze, but the police had only c.e. They were shortsighted not to have gotten a search warrant for his home in Bayonne; there they'd have found fountain-pen/bombs full of acid and other similar devices.

In March, 1934, he went back to sea. His past record of crime, suspected crime and mental disturbances was unknown to both RCA and Ward Line. To them he was just an oddball with a fixed smile, who was an efficient radio officer.

The *Morro Castle* suited him fine: no police investigations. It was good pay, but not good enough. He started pocketing some of the money passengers paid for telegrams, but forgot to doctor the books to hide this. RCA investigated carefully, decided he was a thief and let him know that this voyage from Havana to New York would be his last. So he was on the chopping-block just like Alagna--but he didn't tell him that. As the *Morro Castle* went through the usual predeparture fuss in Havana, Rogers was scheming: buy certain things ... make Alagna seem guilty ... also make the captain and Ward Line grateful to him for revealing this.... The future looked rosy.

He slipped ashore but returned exactly one hour before sailing time. That was what strict regulations required of a radio officer. He was going to be very fussy about official rules. The reason for his last-minute dash ashore was the small parcel he carried up the gangway. It contained two bottles of acid -- sulfuric and nitric.

By the time the pilot had been dropped and the ship was heading north, the pattern of life for the passengers had swung back to normal. It was cocktail time. This was enlivened slightly by the unusual presence of Captain Wilmott. He apologized for his absence from the dining room and used weather conditions as his excuse.

Chief Engineer Abbott was dressing for dinner with his usual care (pomade, stiff white shirt, etc. -- all the frilly items which made Warms's blood pressure rise). He was unhappy. Captain Wilmott had politely suggested that he transfer to the *Oriente*. It seemed as if that worm on the bridge had come out on top. (Actually, "that worm" didn't even know about it yet.)

Warms, meanwhile, was wandering around checking things and had been passing the radio shack when the bulk of George Rogers suddenly filled the doorway. "Mr. Warms, I'd like to see the captain."

"Why?"

"I have some information for him."

"Oh, what's it about?"

"Better if I tell him." (Fixed smile always.)

"OK, I'll get him to see you after dinner."

At 7:30 PM on the dot, as always, Warms and Abbott arrived in the captain's cabin. This pre-dinner meeting was a ritual. The ship's three senior officers discussed the day's run informally. "All OK down below," mumbled the chief engineer. His gruff brevity indicated he was sulking about the transfer -- to which nobody referred. The chief mate reported five stewards logged for drunkenness. No mention was made of unkept fire-drills or the sealed-off smoke-detector system. They did their usual rounds; bridge, prom deck, tea-room. This time, for a change, Wilmott wanted to have a look at the galleys; possible poisoning was still on his mind. Eventually they all wound up in the dining saloon. Wilmott gave passengers the excuse of "pressure of work" for his absence till then, nibbled at a crab hors d'oeuvre, sipped some ice water and excused himself. He had an appointment with Chief Radio Operator Rogers.

"Captain," said Rogers deferentially, "I must tell you about my assistant Alagna."

"Hmm, that S.O.B. What now?"

"I've suspected for weeks that he might stir up trouble, sir. Now I've got proof of his intentions. In his things I found two bottles of dangerous acids."

Wilmott started fuming.

"I suspected the man was crazy! Now I know it! He was trying to start a riot in New York because --"

Rogers was quietly holding up his hand to hush his skipper.

"Don't worry, Captain. Everything is under control. I got rid of them. I flushed them down the toilet."

Wilmott believed him. In contrast to Alagna, Rogers seemed almost saintly.

"Mr. Rogers," he said, "take this key for the emergency room. I don't want it anywhere that man can get it."

He evidently suspected Alagna might try sabotage. It's hard to imagine what harm he could have done in the emergency room, but Wilmott was as neurotic as Alagna on this subject of the radio-compass. He added, "Thank you *very* much for disposing so promptly of those acid bottles."

Should he have suspected that Rogers still had them? Or that they'd be used when the time was ripe?

11:00 PM (Wednesday)

The orchestra in the deck ballroom was playing the last dance. The chief engineer whirled a pretty young female passenger around the floor, then went off to bed. Didn't look as if there'd be anything more than a routine, slightly boring trip to New York. He hoped for a little delay before his transfer.

11:55 PM

Joe Bregstein, passenger, a dentist from New Jersey, was standing on the prom deck, looking at lights on the horizon.

"That's Miami."

The voice behind him startled him. He turned to see the chief radio officer, who (having just been relieved) was heading for bed.

"About 10 miles away. Then we come to Fort Everglades."

"Thank you."

"Good night."

"Good night."

Bregstein watched Rogers walking away, thinking that for a big guy -- mostly blubber -- he moved very quietly.

Midnight (Wednesday/Thursday)

The ship's bell on the bridge struck eight bells. Third Mate Hackney took over the watch from Fourth Mate Hansen. "Barometer is dropping. Looks like bad weather ahead," he was told. Up the long steep stairways from the engine room came the oilers and firemen who'd just been relieved.

In the radio shack, Maki started his four-hour watch.

Far astern, the night watchman was making his rounds, punching time clocks at various places.

1:00 AM (Thursday)

Chief Mate Warms, on the edge of sleep, was thinking about the lack of boat drills. This stuff about not upsetting the passengers ... he'd better get Wilmott to see straight ... he dropped off to sleep.

2:10 AM

Third Mate Hackney telephoned the captain.

"Sorry to wake you, sir, but that tropical storm is almost on us. Shall I --"

"Reduce speed. Zig-zag. Keep our head into the wind. Change every few minutes."

"Right, Captain. Thank you. Good night."

2:15 AM

Chief Engineer Abbott awoke as the ship lurched. He phoned the engine room.

"Everything OK?"

"Yes, Chief. The bridge just ordered reduced speed."

Abbott went back to sleep.

7:55 AM

Rogers relieved Alagna.

9:00 AM

Cruise Director Smith came on the loudspeaker.

"Good morning, everybody. This is your cruise director speaking. Are we happy? ... Everybody? ... Let's hear you! ... Good, good ... Weather doesn't look too good today -- wet. But there's plenty to do: bingo, quoits, miniature horse racing. Ladies, the beauty parlor is open all day. Tonight -- oh, boy -- don't forget, folks, there's a grand elimination dance with lots and lots of prizes."

9:20 AM

Chief Warms, not having seen the captain at breakfast, knocked on the door of his cabin. He had locked himself in. Hearing Warms's voice, he opened the door a few inches and whispered to Warms, "Acid -- that's what they'll use. Acid to destroy me!" Warms wondered if this was symptomatic of a coming breakdown.

10:00 AM

Passenger Joe Bregstein felt quite worried about the lack of boat drills. Meeting Chief Engineer Abbott, he questioned him about it. Abbott was very cheery and reassuring.

"Is this your first trip?"

Bregstein confessed it was.

"Well, there's no need to worry. The *Morro Castle* is the safest ship afloat! I'll tell you what I'll do. This time tomorrow I'll take you on a tour of the ship, top to bottom, and you'll be convinced."

Bregstein thanked him, agreed to come--and bring his young son Mervin, who'd be very interested.

11:00 AM

Alagna was lying in bed, worrying. Rogers had told him "I dumped those bottles overboard. Best thing for you, believe me." Alagna didn't know how they got to his bunk. He didn't know *anything* about them. Was Rogers suggesting that he was helping Alagna keep out of trouble?

Noon

"Eight bells" sounded as Rogers (who'd just been relieved by Maki) appeared on the bridge with a weather report from Miami. There was another force-seven gale coming.

In the dining saloon the gossip was all about Captain Wilmott, who'd become a "monk in retreat" again. At the table where Dr. Van Zile, the ship's surgeon, was host, the general mood was "to heck with it all." They had a big bowl of planter's punch, and were going to make it last right through the afternoon.

12:30 PM

Rogers went to the captain's door to make a report on Alagna. The door was locked. Wilmott told him to go away, and stay away. It now seemed to Rogers that his role as an informer had ended. There would be small hope of keeping his job when they got to New York. Other plans would have to be put in motion.

3:00 PM

The gale hit the ship. Fourth Mate Hansen had just ordered a reduction in speed when Chief Mate Warms appeared in the chart room to see what was going on. Then, surprisingly, Captain Wilmott also appeared. He ordered a further reduction in speed, and stood awhile gazing out into the rain. He went back to his cabin and didn't appear again.

4:00 PM--Midnight

The gale got worse, then eased off. There weren't too many at supper; many passengers were seasick. Smith's projected dance was a fiasco but the weather improved greatly, and when Hansen took over at 8:00 PM he got the OK from the captain (still locked in his cabin) to increase speed.

4:00 AM (Friday, the 7th)

Second Mate Freeman relieved Hackney. "Cape Hatteras is coming up on the port bow," he was told.

In the radio shack Alagna relieved Maki. (Their watch schedules were: 12 to 4, Maki; 4 to 8, Alagna; 8 to 12, Rogers.) "Is all quiet on the 500 kcs front?" he asked. He was referring to the international calling-and-distress frequency, where all ships had to maintain a watch during the Silence Period times.

9:00 AM

First Officer Warms called on Captain Wilmott. Dr. Van Zile had seen him after breakfast because he'd been complaining of headache and tiredness. Warms told him that surveillance of the radio shack was continuing, but that Alagna seemed to be OK.

10:00 AM

Chief Engineer Abbott took Joe Bregstein and his son on the promised tour, to show how safe the ship was. He showed them the button on the bridge that sealed off watertight bulkheads, and the controls on the bridge that showed, by lights, just where a fire, or even a high temperature, could be located, if it arose. The smoke-detecting system also had its rows of lights, but Abbott didn't know Wilmott had ordered it turned off. They looked at the lifeboats, balsa-wood floats, lifebuoys and life-preservers (78 just for children). The climax of the tour was the engine room. Abbott led them down steel ladders, along catwalks, into the hot noisy area of generators and furnaces, talking about electricity, air-conditioning and refrigeration, etc. He hoped he had left Bregstein and son with the impression that everything was under control.

4:00 PM

Cruise Director Smith tried to stir up fun with bingo, a tea dance, horse racing, prize dances, musical chairs. Nothing worked. The previous day's jostling and seasickness had left passengers gloomy.

5:00 PM

Smith made one last effort, announcing over the loudspeakers, "Remember, folks, this is Friday -- the last full day! The captain's farewell dinner and gala ball are tonight. Come one, come all. Wind up with a bang!"

5:15 PM

Captain Wilmott made a supreme effort and forced himself into one last social obligation, his farewell cocktail party. It didn't involve more than a few hand-picked first-class passengers.

7:00 PM

Wilmott joined Warms on the bridge. He saw the weather report. "Reduce speed," he ordered. Then, "I don't feel so good. I'll take an enema and lie down." He left.

7:20 PM

Chief Engineer Abbott was dressing for dinner when his phone rang. He was told there was a fuel blockage in No. 3 boiler. It couldn't be fixed without shutting everything down -- so they'd be unable to get 20 knots for the rest of the voyage. He called the captain. No answer. He called the bridge; was told that the captain wasn't there. Another call to Wilmott got no answer. He called the bridge again, told them about the boiler.

Chief Mate Warms had also been trying to speak to the Old Man. He'd seen him about 15 minutes earlier in his cabin, when Wilmott had told him, "Get back on the bridge. I'll be up soon." He never came. When Abbott had been up for the second time, Warms decided to investigate. He went down to the captain's cabin, knocked -- no answer -- so he pushed open the door.

Horror was waiting for him. He found Captain Robert Wilmott, half dressed, slumped over the side of the bath, his eyes wide open -- obviously dead.

He reached for the phone, called the bridge.

"Hansen? Find Tolman (the purser) and come with him to the captain's cabin. The Old Man's dead."

He closed the captain's eyes. Their sightless staring was unnerving him. Then he called the ship's surgeon.

"Come to the captain's cabin at once, Doctor Van Zile. Bring your bag, please." As he put the phone down, Abbott walked in.

"One of the boilers has gone. I can't get hold of the Old Man. . ."

"He's dead. In there."

Abbott stopped, stared, gaped.

"Maybe he just fell?"

"He's dead. I'm in command. Everyone, including *you*, will take my orders."

Abbott, in a daze, staggered to a corner and flopped down.

Fourth Mate Hansen and Tolman arrived. They were excitedly conjecturing when Dr. Van Zile came in and ordered that Captain Wilmott be lifted onto the bed.

"Look at his color, Doctor. Seems quite blue."

Dr. Van Zile didn't answer Tolman. (Discoloration can come after a heart attack -- or it might indicate poisoning. No way of knowing without an autopsy.) He filled a hypodermic syringe with something and injected it into Wilmott's arm. After doing this several times -- apparently trying to revive the captain -- he finally announced, "The captain is dead."

The purser said to Warms, "That makes you captain. I'll prepare the necessary papers."

Acting-Captain Warms ordered Hansen and the chief engineer to lay out Wilmott's body. Then he turned to the ship's doctor.

"What was the cause of death, Dr. Van Zile?"

"Indigestion and heart failure," he replied.

Strictly speaking, these are symptoms, not causes. He should have said that an autopsy was necessary to be sure of the cause. Warms asked him to get some of the other doctors aboard to confirm his findings. Then he went up to the bridge, to reshuffle the officers' ratings. Second Mate Freeman was to be chief mate; Hackney became second mate. Fourth Mate Hansen was now third. He then gave Purser Tolman a message to give to Rogers, now on watch in the radio shack:

> WARD LINE NEW YORK
> WILMOTT DECEASED 7:45 PM
> ACKNOWLEDGE
>
> WARMS

Captain Robert Wilmott lay resplendent in his uniform, having been laid out by Chief Engineer Abbott and Third Mate Hansen. Hansen saw that Wilmott's face was now turning almost black, and wondered if the doctor's swift diagnosis had been satisfactory.

To back him up, half-a-dozen doctors during the next hour came traipsing into the cabin, led by Dr. Van Zile. All agreed that the discoloration was the after-effect of a severe heart attack. (They were not equipped for an autopsy, anyway.) As they came and went, Abbott was slumped in a corner, silent. Apparently Wilmott's death had induced a trauma in him.

The effect on the rest of the ship's passengers and crew was extreme. Music stopped. The decks were deserted. Some passengers pooled their Cuban rum and made a kind of Irish wake, going round the staterooms. Down below, in the engine room, the removal of that very remote figure, the captain, didn't make any difference to the operations of gauges, compressors, pumps, etc.

The effect on Ward Line VIPs (the telegram was shunted to their various home addresses) was shock, confusion and a reply sent (rather stupidly) over the head of Acting Captain Warms.

ROBERT TOLMAN SHIPS PURSER
SS MORRO CASTLE
PLEASE CONFIRM QUICKLY MESSAGE SENT BY WARMS TO BECKENDORF REGARDING WILMOTT GIVING DETAILS

WARDLINE

Tolman's rejoinder was swift:

CONFIRMING MESSAGE FROM WARMS STOP WILMOTT DECEASED ACUTE INDIGESTION AND HEART ATTACK SEVEN FORTYFIVE THIS EVENING STOP ALL PAPERS FOR ENTRY IN ORDER

TOLMAN

Was Wilmott poisoned? Exhaustive investigations by two people (who never met) say George Rogers killed him. Police Captain Vincent Doyle (whom we'll meet later) reckons Rogers had easy access to the captain's food tray, and simply slipped poison into the coffee or scrambled eggs--as retaliation and fear of Wilmott's evidence against him. Wilmott's lifelong friend, Captain George Seeth, found that Rogers was involved as a key member of a smuggling ring, and Wilmott's knowledge of this made him very dangerous to Rogers. For either reason or possibly both, Rogers couldn't afford to let Wilmott reach New York and blab to the police. This is theorizing, of course, from a remote distance in time with no hope of getting firm evidence.

Let us return to that Friday evening, September 7, 1934, when the *Morro Castle*, with a new captain, is proceeding up the East Coast -- only a few miles and hours away from journey's end in New York.

The weather has turned nasty again. The ship, heading into a gale-force wind, is pitching violently. Warms checks the shore-lights from the wheelhouse. He's very tired. It's been a long, hard night. He is ill-prepared, physically or mentally, to become master of this ship. He has been 24 hours without sleep and Wilmott had never entrusted him (or anyone else) with authority.

Chief Engineer Abbott leaves Captain Wilmott's cabin, goes to his own, undresses, telephones the engine room and is told "All OK down here."

The time is midnight. Saturday, September 8th, is just beginning. Charles Maki relieved Rogers a few minutes ago. Rogers says he will "take a breath of fresh air" and goes out on deck. He's away 20 minutes.

The trick of slipping into the present tense may heighten our attention, but it doesn't help at all to discover what Rogers was up to in that 20 minutes. We can only speculate and juggle with the circum- -(oh, it was going to be shortened to "c.e.", remember?). Here's what was possible and probable.

Twenty minutes were plenty of time for Rogers to make several incendiary devices--the fountain-pen type. He knew well how to make them and had often done so successfully in the past. With the acid he'd brought aboard and some thin strips of copper wire, he had all the raw materials to make these little bombs. The locker in the first-class writing room was the perfect place; it was just below the space where he'd seen sailors (directed by Hansen) move the Lyle gun and the barrel of gunpowder. That gunpowder would make a perfect trailer to spread the fire. Another trailer was the feed line to two gasoline tanks right by the radio shack: all Rogers had to do was to uncouple the feed line, and let the gas trickle along the deck. (Gas can be smelled, and the night watchman did smell it, but he reckoned it was the result of cleaning. He didn't report it.) Once the fire had been well started, it would, in addition to all its other destruction of life and valuables, destroy the evidence of its own birth.

12:15 AM (Saturday, September 8th)

Clarence Hackney, now on watch as second mate, notes that the weather has improved, but that his skipper, Warms, is looking very tired.

"Have a break, sir."

"Later, Clarence, later. Time for sleep when we dock."

"One thing Warms hasn't noticed -- the smoke-detecting apparatus Wilmott switched off for passengers' comfort is still inoperative.

1:00 AM

Night watchmen Pender and Foersch report to the bridge that all is quiet.

2:00 AM

"We're 30 miles south of Scotland Light," reports Hackney to Captain Warms. The ship will soon be in the relative shelter of Ambrose Channel, heading for New York Harbor.

2:45 AM

Stewards Campbell and Ryan are collecting glasses and hinting to partying passengers that it's time for bed.

2:50 AM

"Do you smell anything?" says Ryan to Campbell. They both wrinkle their noses.

2:51 AM

Watchman Foersch reports to Captain Warms that he has seen smoke coming out of the fiddley -- a galvanized iron duct for fresh-air supply to the first-class writing room.

Warms runs to the fiddley, sees the smoke, rouses Chief Mate Freeman ("Ivan! Get up! Fire!") and rushes back to the bridge.

"Clarence! Go below. Find the source; let me know the situation -- FAST!"

Hackney grabs a fire-extinguisher and runs to the deck below. Campbell and Ryan have already located the locker from which smoke was coming, then have run to raise the alarm, passing by a fire extinguisher on the bulkhead. (If they'd seen it and used it immediately, they *might* have made a difference.)

By the time Hackney arrives, three minutes later, the smoke is thick. He opens the locker door and a wall of flame rushes out. He empties his fire-extinguisher into it but nothing can now stop the growing inferno. He remembers that it was he who had put the Lyle gun and its powder keg where they were, but now he forgets that he could isolate the writing room by lowering its fire door. He dials the bridge.

"It's bad! We need water!"

"Break out the hose," calls Warms. "I'll get you the men."

Then, to the engine room, "Bridge here. Captain speaking ... Gimme all the water pressure you can!"

"Why?"

"*Fire!* Want all the pressure possible! And I mean NOW...."

He drops the phone, letting it dangle, not bothering to replace it -- and doesn't hear the engineer below crying, "But Cap'n, we've got a boiler out! We can't give you more than a trickle!"

(News of the faulty boiler had been brought up by Abbott, but had coincided with Wilmott's death, and Warms hadn't heard Abbott. It wouldn't have improved matters if he had heard, nor if he'd listened to the engine room -- but some of his later distress and puzzlement would have been lessened.)

In the present urgency, he turned to an AB. "Wake up the bos'n. Get him and his men to fire stations. HURRY!"

The bos'n isn't just asleep; he is drunk. He can't command anyone or anything. Deck hands are responsible to him, and he to the chief mate (that's the situation on any ship). With the bos'n out of the picture and the chief mate nonexistent (as captain, Warms had no direct contact with the crew), the *Morro Castle* is now supremely disorganized.

Smoke is billowing out of open portholes on the sun deck and the whole ballroom is blazing. It is 2:53 AM -- only a few minutes since the fire was first spotted. *(This incredible speed of the fire was a clear indication of its having been set. The fire was out of control almost from the start.)*

At 2:50 AM Warms makes a bad mistake. To offset the effect of a new gale, he turns the ship *into* the wind; that is a fine textbook change *if there isn't a fire.* As it is, the wind fans the blaze--and this goes on for 12 minutes, from 2:50 to 3:02 AM.

2:51 AM

Warms goes to the fire-detecting apparatus. No lights are on. Reassured, he calls the engine room.

"Any fire down there?"

"No. But a little smoke."

"Have the chief call me as soon as he gets down there," said Warms. Abbott never does get below. He is concentrating on saving his own skin.

Warms looks at the board displaying the fire-extinguishing system. It has begun to glitter like a Christmas tree.

"My God!" cried Warms. "The whole ship's almost ablaze!"

He yanks a lever. Fire alarm bells clang all over the ship. They awake George Alagna. *The three radio operators share a tiny cabin next to the radio shack. "I woke up choking from the smoke," he told the Coast Guard inquiry, weeks later. "I could hear people running, then a scream -- 'The fire hoses won't work -- the pressure is gone!' I ran to wake Rogers, who was still sleeping soundly." (The timing-device didn't need his presence.) "He was a big bear of a man, and it was hard to rouse him. Eventually he got out of bed grunting and pulled on his pants. We both ran to the radio room where Third Operator Maki was on duty."*

Just before they get there, at 2:59 AM, comes a tremendous explosion. The gunpowder stored over the writing room has blown up. Through the ensuing havoc of smashed portholes, a 50-knot wind fans the flames and hastens the death of the ship. This explosion starts a full-scale panic. Passengers stumble around, women screaming, men swearing. The ship is ablaze. Midships, crewmen are hacking at deck chairs and tables with axes -- to throw the pieces overboard as floats -- but there is no officer to direct them.

Alagna arrives at the radio shack, just in front of Rogers. He notes the time. It is 0800 GMT -- exactly 3:00 AM local time.

The light which shows power coming in from the engine room is out. Alagna is switching on the emergency power (lead-acid batteries) as Rogers lumbers in, dismisses Maki and sits down at the console.

"Go to the bridge, George," he tells Alagna. "See what orders the mate has." (Warms, to him, is still the mate.)

Moments later Alagna, white-faced, is back.

"I can't get through," he says. "The flames are taller than the radio shack."

Rogers tries calling on the telephone, then the speaking tube. All he can hear is the roaring noise of the flames.

"Go back!" he orders Alagna. "We've *got* to have an order for an SOS." Alagna flees.

Rogers and Maki peer at each other through the smoke.

"What else can we do?" says Maki.

"Nothing, until orders come, but you can do something for me. Get a sopping wet towel I can breathe through. Maki obediently takes a towel to the wash basin, and Rogers wraps it around his head.

Maki then leaves. *(He will wander around the ship -- now becoming a madhouse of confused people -- and eventually he will jump clear.)*

Out on C deck, Cruise Director Smith finds a group near panic as smoke was swirling overhead. "Down on your knees!" he shouts. "All of you! Grab the waist of the person in front." He gets to the head of the "crocodile" and leads them slowly along the corridor up to B deck. "Get to your lifeboat positions. They'll swing the boats down to you."

Apart from all the confusion, George Rogers sits quietly waiting for orders to transmit an SOS to all ships.

Far below on E deck, Chief Stewardess Lena Schwarz and her colleagues have been running from stateroom to stateroom, awakening passengers, reassuring them, guiding them topside. She hears a cry, "Help me!" Running toward the voice, she eventually looks out of a porthole and sees a woman hanging on the rim of her own porthole, feet dangling over the sea.

"Don't move!" shouts the stewardess. "I'll come and get you."

The ship suddenly lurches, and the woman is flung into the sea. Mrs. Schwarz, seeing the body bobbing in the wake, is tearful, helpless.

At 3:10 AM vital electrical wiring burns through, and the whole ship goes black. The electric steering fails. The ship's rudder won't work. The engineering department switches to emergency power (batteries) that will work the lights for a while. The radio department is already working on emergency batteries. The *Morro Castle* is now drifting, helpless. She is doomed -- every which way.

Still no SOS has been sent.

Alagna is now up on the bridge, but his chance of catching Warms' attention seems zero. The steering gear is useless; Warms is frantically steering the ship by its engines. That way he can zig-zag the ship toward the shore. He himself is zig-zagging on the bridge -- running from one wing to the other, bellowing orders, but deaf and blind to Alagna, who is chasing him back and forth.

Then he begins to order the lifeboats to be lowered; but even this doesn't alert him to the idea of an SOS, nor to Alagna pleading for it. Alagna, despairing again, returns to the radio shack, shedding his cap for a towel as he goes. He enters, muffled-up like Rogers. *("We looked like Bedouins," he will say later.)*

Rogers has news. The R/O of *Andrea Luckenback* has spied flames and asked WSC/Radio Tuckerton about them. WSC (the main New Jersey coast station) is now asking CQ (all ships) for any info about a ship afire.

Rogers says, "I'm itching to reply, but we have no orders!"

"Here's what we'll do," says Alagna. "The Silence Period starts in a minute. Get on the air, tell everyone to stand by. That'll get their attention. I'll go back to the bridge again. Goddamnit, I have been up there three? -- or is it four? -- times already! I'll *force* the Old Man to OK an SOS."

Rogers agrees. They watch the second hand of the big GMT clock advancing by little jumps toward 0815 ... the second hand comes to 12. Rogers hits the key.

CQ CQ CQ de KGOV KGOV ... but the powerful signal of WSC breaks in.

SP! QRT (stop sending).

Rogers looks at Alagna.

"WSC says stop."

"Stop nothing. This is an emergency. Keep 'em waiting. I'm off to the bridge to get some action." He disappears.

Rogers lifts his feet from the hot floor of the radio shack onto the rungs of his chair. He maneuvers his blubbery bulk sideways and crouches over the Morse key.

EMERGENCY CQ CQ CQ de KGOV KGOV KGOV PSE QRX ("please stand by for a message soon on...") EMERGENCY

Up on the bridge, two decks above him, Warms has discovered Abbott, lurking in a corner of the chart room.

"What the hell are you doing here?" he yells. "You should be down in the engine room to get my orders!"

"It's too late...," Abbott begins.

"Damn you! Get below! Organize things! We need *water*!"

The calm quiet voice of Hackney intervenes:

"Captain, the water pressure's gone."

"Gone?" shouts Warms. He whips around on the chief engineer again. "What happened to the water pressure?" Abbott is silent. "What happened to it? Answer me. Answer me!" (He has never, in the excitement of Wilmott's death, understood about the boiler failure.)

Alagna has arrived on the bridge again. *(His own words at the Coast Guard hearing describe what happened next.)*

"Q. When did you finally get his attention?"

"A. I pleaded with him for some time, telling him 'We've got to send an SOS.' But he kept running over to the port side to take another look at the fire. I couldn't understand why he didn't look at the starboard side where people were jumping over and where there was so much to be done. I followed him around, seized his sleeve and finally got him to understand that the radio room was taking fire. 'Rogers is dying down there!' I yelled. 'It'll soon be impossible to send a message.' "

"Q. What happened then?"

"A. Captain Warms was muttering, 'I can't help it. I can't do anything.' He looked shattered. I said, 'How about some orders?' He looked at me as if some inspiration had come to him and said, 'Can you send an SOS?' 'YES YES YES!' I nearly screamed, 'I've been trying to tell you that all along!' 'Then send it,' he said. I asked him our location and he said, 'About 20 miles south of Scotland light vessel.'"

The time is exactly 3:18 AM (0818 GMT). An SOS should have been sent around 0300 -- when the gravity of the situation made it imperative. It takes Alagna a long seven minutes to return to the radio shack, as smoke and fire bar his way.

While he's fighting his way back, we have time at last to pick up the interrupted tale of Chief Engineer Abbott. When he'd been awakened by the fire bells, he went next door to the first assistant's cabin. "Is there a fire?" he asks. First Assistant Bujia doesn't answer but puts on his work clothes and goes down to the engine room. Abbott returns to his cabin and changes into his dress uniform. He phones the engine room; the first assistant says they are trying to supply water for the fire fighters. The smoke is already coming down into the engine room.

"Do the best you can," says Abbott. He finishes dressing, puts on his gold-braided cap and goes out -- but not down to the bowels of the ship. He watches a fire crew trying to couple a hose to a hydrant. They can't get any pressure. He watches seamen throwing chairs over the side to provide things people could cling to in the water. He walks down to C deck where a group of girls is standing. "What should we do?" they ask him.

He tells them to wait on the boat deck. He goes down to D deck. There he meets First Assistant Bujia coming up. "Where are you going?" he asks. "To the bridge," says Bujia. "I called you. They're managing things OK down below, but they can't last there much longer."

Abbott stares at his subordinate. His (Abbott's) duty is clear. He should go below himself to check out the situation. Instead: "Go back. Stand by. I'm going to the bridge," Abbott tells Bujia.

He'll regret those words for the rest of his life.

Alagna bursts into the radio shack yelling "OK! Send it! We're 20 miles south of the lightship, at Sea Girt!"

It is 3:25 AM. Rogers, who has been waiting for this for 25 minutes, goes through the proper procedure: by the strict regulations which had accumulated since the *Titanic* disaster, he has certain things to do *before* sending the SOS itself.

He sends a four-second dash (DAAAAAAAAAAH) then takes his hand off the key for a second. DAAAAAAAAAAH again for four seconds, silence for one second. This sequence he continues for a minute--sending 12 DAAAAAAAAAAHs and 12 pauses in between. This should be more than sufficient to operate the auto-alarm bells of every ship within range.

Waiting then for another minute (to give the roused Sparkses on all the ships time to get on watch and tune in) -- he sends the message at 3:27 AM.

> SOS SOS SOS DE KGOV: MORRO CASTLE AFIRE OFF SEAGIRT 20 MILES SOUTH OF SCOTLAND LIGHT VESSEL NEED IMMEDIATE ASSISTANCE
>
> MASTER

The nearby *Andrea Luckenback* got the message intact. The *Monarch of Bermuda* asks for a repeat of the position. Several other ships acknowledge the message.

Rogers becomes conscious of his feet being very hot. His own account of what happened next puts him in a heroic aura. *("I had a wet towel over my face and I could hardly breathe. I'd gotten about halfway through the distress message when the corner of the radio-room table housing the receiver batteries exploded. The room filled with sulfuric gas. The receiver was out of commission but I continued sending the SOS as the transmitter was still OK. After sending the SOS the auxiliary generator stopped.")*

He staggers over to the emergency generator room, connects the wires that had come adrift and gets the generator going again.

("I was just staggering around. My feet were burning bad. I remember the wet towel on my face but it was permeated with smoke and I couldn't hold out much longer. Then there was an explosion in the generator room and this time the generator stopped completely. I felt Alagna shaking me. He was saying, 'Come on, Chief, don't you want to see your wife again? Let's get out of here.' ")

But Rogers is singleminded -- blinkered like a horse to see one way only.

"Go back to the bridge. See if there's anything else." This will be the fifth time Alagna had been up there. He has no trouble finding and speaking to Warms this time.

"Evacuate the radio shack! We're abandoning ship!" said Warms.

Alagna finds Rogers sending another message.

SOS CQ DE KGOV 20 MILES SOUTH OF SCOTLAND LIGHT STOP CANNOT WORK MUCH LONGER STOP FIRE DIRECTLY UNDER RADIO SHACK NEED ASSISTANCE IMMEDIATELY

Before Rogers can send a signature, the generator explodes again. He collapses on the table, head down, sleepy, thinking (he said later) that if this is death, it doesn't hurt much.

"This place won't last another two minutes!" Alagna shouts. "Skipper says abandon ship! Let's go!"

Rogers struggles to his feet. With Alagna pushing him, he oozes his 250 pounds through the passageway. Out on the open deck, he looks around at the fire.

"We've nothing to be ashamed of," he says. "We waited for the SOS and sent it."

Out by Number Two lifeboat, Eban Abbott is ordering "Lower away." There are only eight people -- six of them crew -- in the boat with a capacity of 70.

"*Don't* lower that boat!" yells Warms from the bridge, "keep it at the rail for passengers."

Abbott's counterorders keep the boat going down. It slithers down on its davits. Warms said later, "It was a moment of shame for all who believe in the tradition of the sea." It was, indeed.

When the *Monarch of Bermuda* gets the distress message at 3:27 AM, her captain, Albert R. Francis, shoots off a volley of orders -- to raise steam, prepare lifeboats, set up casualty clearing stations, get coffee, tea, soup, rum handy. When his ship is alongside the *Morro Castle* at 7:00 AM, they are ready for anything. Five lifeboats are soon bobbing toward the burning ship. This seamanship receives the admiration of Alagna and Second Mate Hackney for "the calmness and control you only get on a good British ship."

Meanwhile, George Rogers is busy supplementing his firmly established role as the hero. Twice he tries to get through a wall of flame to rescue passengers. His most theatrical performance is his attempt to rescue a woman passenger who'd become stuck in a porthole.

Getting two seamen to hold his feet, he dangles upside down trying to reach her. He fails, is hauled back on board, and is going to try again, swinging on a rope like Tarzan, but unfortunately the screaming woman frees herself, falls into the sea and eventually drowns.

He has made vivid impressions on many people. If RCA or Ward Line -- or anyone -- tries to revive old ideas about him, how ridiculous they will seem.

There is now a general alert along 300 miles of the coast. The Coast Guard Patrol Boat *Cahoone* gets the order to speed to the rescue at 4:00 AM. The U.S.C.G. cutter *Tampa* is delayed until 5:30 AM. The only Coast Guard surface boat that is on the scene fairly soon puts out from Sea Girt to investigate the "ball of flames" they see. What they see when they get up close is, according to their helmsman William Moulton, something he never wanted to see or hear again. "The ocean was alive with screaming men and women and so many grabbed the surfboat it nearly capsized. The crew did all they could to keep others off. We had to be very careful not to run over someone in the water. All around were men and women calling for help and we, already overloaded, were unable to help at all. They were piled three deep over the engine, boxwood-fashion. Just how many I didn't know."

"This way, Chief," says Alagna, going to the deck below. From there they scramble down another 20 feet to join Captain Warms and a dozen other officers and crew on the bow of the ship. This place is, at the moment, an island of safety. The only other place reasonably safe from the fire is at the other end of the ship, where passengers are crowding to the stern rails. In various parts of the ship they are pouring through passageways, screaming at portholes for help, flopping into the sea or shinnying down ropes -- no one to command them. Panic rules. Some half-clothed passengers stagger out into the rain, and see members of the crew in a lifeboat, rowing away as hard as they can. In between is the flaming superstructure. From some portholes arms are waving desperately; these people cannot be approached -- and they have no escape.

Passenger panic is understandable, forgivable. Panic in the crew has no excuse. Chief Engineer Abbott, in full dress uniform, takes off in Lifeboat Three, with 16 crew and no passengers. Before they reach shore (by which time there will be 29 crewmen plus 3 passengers they'll pick up), Abbott has torn off his epaulettes and all signs of rank, presumably to be anonymous.

Not all the crew are part of a shameful story. Dr. Van Zile's body will be picked up by a fishing boat; he dies trying to save women and children.

The *Andrea Luckenback* arrives and flashes blinker signals at the men on the bow:

DO YOU NEED ASSISTANCE?

A foolish and unnecessary request? No, the ancient laws of salvage state that the vessel in distress has to ask for help. The *Morro Castle* blinker is still on the bridge, but Warms has a flashlight. "Here, Sparks," he says, handing it to Rogers, then dictates to him:

YES IMMEDIATELY PLEASE HUNDREDS OF PASSENGERS ASTERN.

The Luckenback answers:

WILL SEND BOATS.

They do that and take off their first load of passengers. The *Monarch of Bermuda* comes alongside next. They are so close that Warms bawls at them through his cupped hands: "Take 'em off aft! We're OK here!"

The *Monarch of Bermuda* picks up 71; the City of Savannah gets 65; the *Andrea Luckenback* 26. That makes 162 in all.

The *Morro Castle* is drifting with groups of people at either end. The whole midships is ablaze and smoking.

On the forepeak, the crewmen and officers are watching the U.S. Coast Guard cutter *Tampa* approaching and they are talking about the fire.

"How come it got a hold so quickly?"

"Must have been set, to spread so fast."

"I think it had a chemical origin."

"What do you think, Sparks?" says Hackney. Rogers is sitting beside Alagna.

"Guess I was too busy getting off the SOS to notice the color of the flames."

Captain Warms doesn't have any suspicions about Rogers, but he does about Alagna, who's been the scapegoat for some time. For now he is just concerned with getting any orders from Ward

Line. "Rogers," he orders, "ask the *City of Savannah* to contact Ward Line and get some orders."

Rogers, however, can't get the attention of that ship nor any other; they are all too busy rescuing people in the water.

The *Tampa* arrives at long last. They offer to tow the *Morro Castle* to New York, if they can stay afloat. Warms, figuring that Uncle Sam won't claim salvage money, consents. The *Tampa* heads for New York harbor, towing the *Morro Castle*. News of the disaster brings thousands of people to the waterfront. Ward Line offices are besieged by newsmen and relatives of passengers and crew. Newspaper headlines keep track of the death toll -- it will be 100 by nightfall but still 34 short of the final figure.

Later that day, at noon, the captain and crew of the *Morro Castle* are on the *Tampa* wrapped in blankets, sipping coffee. The pharmacist aboard diagnoses Rogers as "suffering from acute nervous exhaustion and inhalation of smoke." He is put to bed and sleeps through the afternoon.

Warms, now no longer even a temporary captain, watches his ship being towed a short distance behind. By 6:00 PM they are abeam of Asbury Park, when the hawser linking the two ships snaps. The *Morro Castle* is carried on by momentum, and runs aground right in front of radio station WCAP. (There she will lie for a week, attracting sightseers from all over. Signs posted -- TWO MILES TO THE MORRO CASTLE WRECK -- 25¢ TO SEE THE MORRO CASTLE -- BENEFIT FAMILIES OF THE DEAD. Newsmen paid $5 to board. (A gas mask cost $5, a flashlight $1.)

George Rogers is the last survivor to be taken off the *Tampa* on a stretcher to an ambulance. "Who's that?" asks the reporter. "That's the hero of the day," says Second Mate Hackney. This is like the scent of a fox to the newshounds. They chase the ambulance to the Marine Hospital, and find Rogers in a private room, his wife by his side. (His marriage had been a calamity from the start, but this was a special occasion.) He poses for the photographers. In the limelight he can afford to be modest. "I only did what anyone else would have done."

Warms was blamed, at the Coast Guard hearing, for the 18-minute delay in sending the SOS. Also for driving the ship for 12 minutes at 20 knots into a 20-knot wind (making 40 knots to fan the flames) instead of slowing down. He was indicted for a number of other things.

Chief Engineer Abbott was charged with failure to assign members of his department to their stations during the fire, failure to report to his own station at that time, doing nothing to increase water pressure, nothing about the ship's lighting and generator, and for making no effort to rescue anyone else when leaving the ship in a lifeboat.

In the next few days, U.S. Department of Commerce hearings, legal arguments, newspaper interviews and editorials kept the pot of charges and countercharges boiling. Rogers swung this way and that, as he saw fit. When Alagna hinted to Damon Runyon (who was then a reporter for the New York *America*) that a firebug had been aboard, there was consternation all around, and Rogers got uneasy.

As the official investigation continued, he decided attack was the best defense. On the stand he labelled Alagna as an "agitator," a "vengeful person," "radical troublemaker." Alagna was arrested the next day and put in the House of Detention in New York. There he stayed while the hearings went on for weeks.

Many things came to light: the lack of fire drills; the closing down of the fire and smoke detectors to prevent unpleasant smells for the passengers (both Captain Wilmott's fault); Warms's previous demotions for lack of fire drills; Abbott's shameful stripping off his insignia in the lifeboat.

When Rogers took the stand again, he caused two sensations. The first was his statement that the SOS should have been sent 45 minutes earlier. (This was an exaggeration, stretching 25 minutes into 45.) His second statement, equally sensational, took all the heat off Alagna. He now retracted his previous suspicions and direct attacks. He went to the other extreme, with praise for Alagna.

"I want it to be known that the earlier testimony at this hearing was given with the greatest reluctance on my part and had no bearing on George Alagna's conduct or his responsibility to me, who was his superior on board."

So Alagna was freed from custody. He was going to have many more "downs" in his life before the "ups" won. At one stage he even tried suicide. Later he fell in love and eventually had a happy marriage.

Rogers, for a while, was on the crest of a wave of popularity. RCA withdrew its dismissal notice. The Veteran Wireless Operators Association gave him a medal. He got $500 for just appearing on the stage in a local New Jersey theater. Posters shouted, IN PERSON! RADIO HERO ROGERS TELLS INSIDE STORY OF MORRO CASTLE DISASTER. He received a dozen offers of employment with various shipping lines. He turned them down, saying his wife wanted him home in Bayonne.

He made numerous public appearances. One of these was an official reception given by the mayor of Bayonne. One of the dignitaries there was Bayonne police officer Vincent Doyle. He was made suspicious by Rogers's braggadocio and some doubtful claims he'd been making -- most of which turned out to be outright lies. Doyle had been a ship's radio officer himself. Since coming ashore, he had earned an international reputation by installing one of the first two-way radio systems for the Bayonne Police Department. He knew about radio matters.

"I welcomed you here as the hero of the *Morro Castle*," he told Rogers, "but the newspapers quoted you as saying you were dragged out of the radio room when it was so hot that solder was melting on the terminals of your transmitter. Was that true?"

Rogers just glared, didn't answer.

"I'm sure," continued Doyle, "you've used a soldering iron many times. You know how hot it must get before solder will melt. Have you ever tried to hold an iron that hot in your hands?"

Rogers continued his silent glare.

"You've had a trying experience," Doyle said. "My confidence, however, will not allow me to call you a hero. A hero should be modest and truthful. You are neither. I feel sorry for you. Good night."

That was his first contact with Rogers. He was to become a very important person in the story of Rogers, who'd try to murder him, but let's tie up some loose ends in the *Morro Castle* story.

If convicted, Warms and Abbott faced prison sentences up to 10 years, but for the meantime they were released on bail bonds of $2500 each. Then, by legal maneuvers, their actual trials and sentencings were put off and put off. For insurance, a game of legal bluff went on for a while, involving Lloyds of London and claims amounting to over a million dollars. Lloyds had insured the *Morro Castle* for $45,000 -- and that, in the end, was almost exactly what was collected. The *Morro Castle* was sold as scrap for $44,065.

On the very day she was towed away from Asbury Park, Rogers opened a radio shop in Bayonne. He was soon telling his customers how lucky they were to have him, the *Morro Castle* hero, mend their radios.

He'd been established there less than two months and was doing very badly, businesswise. He had just left the shop "to get a breath of fresh air" when it caught fire. His alibi was unshakable, as usual. Arson was suspected, but "c.e." was the only stuff available -- as usual. He collected the insurance of $1200.

A year later in early 1936, Warms and Abbott were put on trial -- and again Rogers was in the headlines, giving evidence. Warms got a sentence of two years in prison, Abbott four years. Both appealed, and again they got a delay.

In April, 1937, the Court of Appeals, in a strange, unanimous decision, whitewashed both Warms and Abbott. Both their convictions were set aside. Warms "maintained the best traditions of the sea by remaining on his vessel" was one verdict; and Abbott was excused for his actions because he had been "suffering from smoke and was not responsible."

Rogers then met a Bayonne businessman who'd been at that reception where Police Chief Doyle had called Rogers a liar. He was sympathetic to Rogers and smoothed things over between Rogers and Doyle. Rogers, in the end, was actually in the police department, helping Doyle!

It was a strange situation, in several ways. Doyle was apparently ready to forgive Rogers; the police didn't check into his record as a repeatedly suspected arsonist; and he wasn't given any physical examination.

The two men did have some common grounds -- seagoing and electricity. Rogers's arrogance was always an obstacle, forever telling Doyle how clever he was. He was preoccupied with exploding devices. Once he went into a lengthy explanation of how fires could be triggered by timing devices.

"Is that how it was on the *Morro Castle*?" asked Doyle quietly. Rogers just looked at him and smiled.

Doyle began making a record of Rogers's statements. Finally, after nearly a year, on March 3, 1937, Rogers was bubbling away about fountain pen incendiaries when he saw Doyle gaping at him. He'd obviously gone too far. They became silent, hesitating to look at each other.

Doyle thought long and hard about how to reveal his sensational evidence. The next day as he came to work, he met Rogers at the door.

"There's a package for you," said Rogers.

On the workbench was a parcel. Doyle unwrapped it to find a fish-tank heater. Nothing unusual so far. Doyle often repaired electrical equipment like this for his colleagues. The note attached to it said, "Please install the switch in the line cord and see if the heater will work. It should get slightly warm." It had no signature, but Doyle saw it had been typed on the radio-room typewriter. He recognized a faulty letter on the font. He turned to ask Rogers for a comment, but he had disappeared. Doyle plugged in the heater and switched on the power.

The immediate explosion (big enough to shake the main police headquarters a hundred yards away) smashed Doyle's left hand, left leg and right foot. His eardrum was also damaged. He was rushed to Bayonne Hospital, had emergency surgery, and was hospitalized for 18 weeks.

Rogers was now suspect on several counts. Through the c.e. painstakingly assembled by Doyle he could be accused of starting the fire on the *Morro Castle*. Or he could be accused of having attempted to murder Doyle. Another aspect would have been his mental state. Suppressed sexual desire leading to a criminal behavior might have justified a plea of insanity; there was plenty of medical evidence. The decision was to concentrate on the attempted murder of Doyle.

On December 15, 1938, he was sentenced to serve from 12 to 30 years in Trenton State Prison. "Your crime," said the judge, "is of the most diabolical nature. It fell short of murder only by the intervention of Divine Providence."

After nearly four years in prison, he was granted a parole to "join the armed services." It was November 24, 1942, and the United States had been at war for nearly a year. The Navy wouldn't accept him, but the FCC gave him a 90-day permit to operate as a radio officer at sea. RCA assigned him to a ship going from San Francisco to Australia. In Darwin he was arrested -- for what crime it's not clear--but this dovetails neatly with every brush he'd had with the law since the age of 12; thick suspicion, then evidence -- and nearly all of that was c.e.

So he came back to New Jersey and worked in a war plant in Jersey City. He was dismissed, suspected of stealing. He went to another plant in Brooklyn. Again he was dismissed, again on a suspicion--this time of poisoning the water cooler; *again* it was c.e.!

Back to Bayonne. Back to the radio repair shop. Back to financial difficulties. The year now is 1952. He got chummy with an 83-year-old retired printer called William Hummel and his unmarried daughter Edith. Hummel had a similar interest in electrical gadgets and was a sympathetic listener to Roger's latest complaint -- that he'd been framed by the Bayonne Police Department and Doyle because they'd been jealous of him -- the hero of the *Morro Castle* disaster. Even after all those years they were "keeping customers away from the shop," said Rogers. Hummel lent him money. By June, 1953, this amounted to $7500. Hummel wrote in his diary, "It has been unwise on my part to advance Rogers money. He is very temperamental and I'm in constant fear he might renege. I'd lose everything."

He put up his house for sale, planning to move to Florida. His diary had a note, "Must collect from Rogers." This note was discovered by the police when they broke into his home on a tipoff. Hummel and his daughter had been bludgeoned to death.

The police took a long time collecting evidence. They wanted to be sure this time. Rogers wasn't put on trial until 15 months later, on September 13, 1954. The painstaking investigation paid off. In a short time the jury found him guilty of murder in the first degree and recommended life imprisonment.

With Rogers safely tucked away in prison, and only a short time to live, let's have a condensed review, and add a few things not mentioned so far. When arson is suspected, experts ask: Does the fire reach great intensity before discovery? Does it spread rapidly? In an unusual manner? (Above all) do flames change color when water is applied? Answers for Morro Castle are yes, yes, yes and yes.

<u>Incendiary Pens</u> -- Rogers's specialty; described by him with great detail to many people.

<u>M.F.'s (Mysterious Fires) in Rogers's Lifetime</u>

1917 "Black Tom explosion" (Brooklyn Navy Yard in World War I. From this he probably got blueprint ideas for arson. He wasn't in any way connected with this fire, but he was fascinated by it.)

1920 M.F. at his New York job.

1929 M.F. at his New York job. (Egert Wireless Co.)

1934 The Morro Castle (incineration pen in writing room, timed for 2:30 AM)

1935 Radio repair shop, Bayonne.

<u>"Trailers" to feed Morro Castle fire</u>

Gas tanks, uncoupled in a few minutes. (Watchman smelled, but didn't report.) Another has evidence about it in radio log note made by Maki at 2:20 AM. He saw smoke curling from behind wastebasket, portside radio shack.

<u>Unproven Crimes</u> (C.E. -- circumstantial evidence -- jailed Rogers twice.)

1911 to 1914 Petty thefts

1920)

1929) M.F.'s mentioned

1934)

1935)

1937 Attempted murder of Doyle

1943 Trouble in Australia

1944 Suspected in New Jersey of stealing

1945 Suspected in Brooklyn of poisoning water

1953 Murder of Hummel and daughter

In 1956 he was asked point-blank: "Did you start the *Morro Castle* fire?" His eyes glittered and shifted, going every which way -- up, down, around.

"If my case is brought before the Court of Last Resorts, I'll tell everything."

That cryptic answer might be regarded as a preparation for a confession of guilt, but, as with everything else connected with him, we'll never know for sure.

Two years later, in January, 1958, he died of a stroke in Trenton Prison. He was 56.

THE STORMY WORLD AND HEROES BEHIND A QUIET LITTLE MEMORIAL

Battery Park, at the base of Manhattan Island, is a quiet place. The contrast it makes with the furious weekday activity of Wall Street and with the twin towers of the World Trade Center (both of these places not much more than a stone's throw away) is emphasized -- almost epitomized -- by the small (10-feet-high) memorial to wireless operators lost at sea. The original memorial and its nearby fountain, both made of granite, had been erected in 1915. The present structures replaced these in 1954, when there was a relocation and rededication. New York City has commissioned the Veteran Wireless Operators' Association (VWOA) to maintain them. They are so quiet and apparently insignificant that most tourists miss them. (A cop whose beat included Battery Park didn't know about the memorial for years!) The original dedication ceremony on May 15, 1915, has been echoed by a memorial ceremony in mid-May every year since then.

Looking at the photos, you'll probably agree that the memorial contrasts well with the surrounding skyscrapers, especially with the World Trade Center, soaring up through a framework of trees,

BATTERY PARK –
BASE OF MANHATTAN
ISLAND, NEW YORK

Looking Northwest, to the Twin Towers of the Trade Center.

Looking North

MEMORIAL TO
WIRELESS OPERATORS
LOST AT SEA

Looking Northeast

Rear View

ERECTED IN MEMORY OF WIRELESS OPERATORS LOST AT SEA AT THE POST OF DUTY.

JACK PHILLIPS
SS TITANIC
APRIL 15, 1912
ATLANTIC OCEAN

44 NAMES
(AS OF MAY 1989)
+ 2 PLAQUES
OF WW I AND
WW II HEROES

Looking Southeast, at the more impressive monuments to the dead heroes of the US Navy (not Merchant Seamen).

(Photos by author)

and with the much more imposing monuments to U.S. Navy heroes on the other side. The *New York Times* had proposed the memorial in 1912 to perpetuate the memory of Jack Phillips, the 25-year-old chief operator of the *Titanic*, and donated $500 toward this. Harold Bride (Jack's buddy, second operator) gave $25. The Marconi Company of New York contributed $5,000 -- and turned out in full force (leaving their office empty) at the original ceremony.

The original purpose of the monument was enlarged to include all future wireless operators who would die trying to get help for their shipmates. A couple of them actually preceded Phillips in 1909 and 1910; this is why Phillips's plaque is a little below the top left-hand corner. The base plate at the rear of the monument needs a little explanation. The VWOA came into existence in 1925, and has been gung-ho ever since in its aims of fostering esprit de corps among radio men, acquainting the public with traditions and ideals, and in recognizing meritorious services by awarding testimonials, medals, etc. The upkeep of the memorial and the casting of bronze plaques to heroes is a proud duty of the VWOA; we'll return to those bronze plaques presently. In the early days of VWOA, there was an extra wreath-laying ceremony on a cold winter's day -- December, 1927 -- attended by Guglielmo Marconi himself and his one-time employee, David Sarnoff, at that time the head of RCA. Sarnoff (who had been hired as an office boy at the age of 15 for $5.50 a week) remembered the *Titanic* disaster vividly. As a Marconi operator, he had manned his key (atop the Wanamaker Building in New York) for 72 hours straight, receiving, transmitting and relaying messages. Honorary presidents of VWOA include Herbert Hoover 1929-1937, Marconi (shortly before his death in 1937), Dr. Lee De Forest ("Mr. Radio") 1939-1961, David Sarnoff 1962-1971, Barry Goldwater 1975 to the present.

"Meritorious services" does not necessarily have to mean self-sacrifice. The idea behind the use of radio at sea is to get help when needed. This has been accomplished again and again without the loss of the radioman's life. The first time this happened, Jack Binns on the S.S. *Republic* in 1909 saved hundreds of lives on two ships by his devotion to duty. He was an unwilling, embarrassed hero, feted and even mobbed both in New York and his home town in England. He paid the price of being the "first real wireless operator hero." He was also the first name on the Marconi Scroll of Honor. This list (44 as of 1987) has been added to periodically by the VWOA, under the heading of "award of recognition of outstanding acts in line of duty." (Those who earn this and lose their lives doing so are also remembered by a plaque on the Battery Park Memorial.) While still on the subject of awards, mention must be made of the highest possible -- the equivalent of a Medal of Honor (USA) or Victoria Cross (UK). This is the Medal of Valor.

So the VWOA has three ways of recognizing meritorious service: (1) putting the name of the Sparks on the Marconi Memorial Scroll; (2) adding that name to those on the Battery Park Memorial (which means that Sparks died in a shipwreck, trying to get help); (3) awarding the Medal of Valor for either of the above.

Before discussing some of the bronze plaques on the memorial, we must admit that the list gives a very sketchy, inadequate picture. Of all the bronze plaques in Battery Park, we'll select just a few, and tell the story behind each, with a brief biography. These names are part of a very different world from the one in the park where pigeons mutter and the wind gently fans the overhanging leaves. They belong to a world where the wind roars, mountainous seas swamp a sinking ship, where lifeboats are crushed and death is everywhere -- while the radioman is pounding away at his Morse key (... --- ...) in his attempts to get help. That's the world we glimpse in these few background stories.

GEORGE ECCLES (31) (Am) ss Ohio Aug 26 1909

The ship struck a rock at 1 AM. Eccles, in his pajamas, sent CQD (the precursor of SOS) again and again. There was no answer from anyone. He helped the purser bring up a sick soldier from a stateroom below, then the purser was trapped by a falling beam. Eccles tried but couldn't free him, then went out on deck, where he found that 200 passengers and crew had taken to the boats. He was then killed by falling wreckage. He, the purser, the sick soldier and a quartermaster were the only fatalities. His CQD didn't save them, but it saved everyone else. Alert radio operators at the Ketchikan (Alaska) radio station relayed his CQD to two steamers and they rescued all 200 survivors.

JACK PHILLIPS (25) (Brit) ss Titanic Apr 15 1912

His story is well-covered in the chapter on the *Titanic*.

FERDINAND KUEHN (21) (Am) ss Monroe Jan 30 1914

The ship collided with another. Kuehn sent a hasty SOS (heard only by a coast station) just before the engine room filled with water and power was cut off. He surrendered his life belt to a woman passenger without one. Then he slipped on deck, fell into the icy sea and drowned. He and 40 others perished.

GEORGE ARUNDEL GEARE (20) (Am) Mango Reva Fall 1916

He had come to sea to rest his eyes before entering Cornell. He sent an SOS ... then "ALL BOATS GONE." Two ships came but found nothing, nobody. No remains of the ship were ever found.

JOHN FRANTSEN (27) (Norwgn) ss Grontoft March 1922

The audience consisted of stunned, helpless wireless operators on dozens of ships scattered all across the storm-tossed North Atlantic; they were listening to the passing of a great soul. The main character in this drama was John Frantsen, calling out as his ship went down beneath him. He had sent a number of smooth, unhurried--almost apologetic--SOSs; in a few hours he must have known that his situation was hopeless: his ship was in a broiling hell with no one near enough to help. Even so, every time he touched the key he had some joke to make. One of his last remarks was, "Where is my hat?" Eight hours after his last message, the first would-be rescue ship arrived. There was no trace of ship or any survivors.

FRANK CALDWELL (22) (Am) ss Iowa Jan 11 1936

A ferocious storm lashed the Pacific Northwest, and washed the ship ashore. It was being pounded to pieces when Caldwell sent out his SOS. A Coast Guard cutter tried to reach the stricken ship but was unable to launch small boats. The would-be rescuers had to stand by and hope for the weather to abate. It was not to be. The *Iowa* is listed in the records as "lost with all hands." Perhaps the most tragic irony is that Caldwell died within sight of his own home, and his mother had stayed on the beach until all hope was gone.

PAPAS THEODOROU (39) (Greek) ss Kyleene Ap 11 1938

The Marconi Medal of Valor was bestowed posthumously by the VWOA to the Greek consul in New York for delivery to his next of kin. His dramatic history was summed up by a New York *World-Telegram* headline: HERO, TOO BUSY TO GET LIFEBELT, DIES AS 29 ARE SAVED AT SEA. An explosion tore the ship in two, 200 miles off the Azores. Three men were killed or washed overboard. The wireless transmitter was smashed. Theodorou rigged up an emergency set. He tapped out calls for help throughout the day. Three tankers, answering, got there as darkness closed in. All men were taken off -- except Theodorou! He had stuck to his post until the last man had been rescued. Then he jumped for a lifeboat in the dark and was drowned. If he hadn't been "too busy to get a lifebelt," he'd have probably survived.

DONALD MACNEIL (18) (Brit) ss Tresillian Nov 30 1954

The ship capsized and sank off the southern coast of Ireland, carrying a cargo of grain from Canada. At 5:30 AM, MacNeil reported that the ship had developed a heavy list to port, and all the port-side lifeboats had been washed away. *Tresillian* turned over and sank at 8 AM (only two and a half hours from beginning to end), but until then MacNeil had been exchanging with other ships and sending signals for them to take radio bearings. MacNeil's body was picked up with several others. The radio operators of a British coast station paid a spontaneous tribute to his performance, mentioning his unparalleled standard of operating and unfailing courtesy even to his last signal.

WILHELM SIEMERS
1905-1957

WILHELM SIEMERS (52) S.S. *Pamir*, September 21, 1957. The ship (German) was a four-masted bark of only 3,100 tons, leaving Buenos Aires on a training cruise with a crew of 86. When hurricane Carrie hit the ship, the grain cargo shifted. Soon the vessel was in severe distress. R/O Siemers communicated with four other vessels. Within an hour he was sending HURRY -- SHIP TAKING ON WATER -- IN DANGER OF SINKING. He waited too long before abandoning the radio shack. He was trapped as he tried to jump clear. He was one of 80 deaths. Only six survived.

CARL JOHAN NIELSEN DEJLIGBJERG
1901-1959

CARL DEJLIGBJERG (58) S.S. *Hans Hedtoft*, January 30, 1959. The northern seas between Scandinavia and the ice-capped island of Greenland ... blinding snowstorm ... hurricane winds through gloomy days and long nights of the Arctic winter. Into this scene comes from one direction the homebound *Hans Hedtoft*, a joyful boatload of people expecting a delayed New Year's welcome back in Denmark and keeping their Sparks very busy with telegrams; from the other direction, an iceberg. Carl sent out the grim telegram to all ships: "COLLISION WITH ICEBERG ENGINE ROOM LEAKING."

Danish and German ships responded; also a U.S. Coast Guard cutter. Later (no one having yet arrived) GOING DOWN SLOWLY NEED IMMEDIATE ATTENTION made ships and planes redouble their efforts. A whole day went by. A German trawler got to the position indicated but could see nothing through the thick snow. Carl radioed VERY MUCH ICE EVERYWHERE. GOING DOWN SLOWLY. MUCH WATER POURING INTO THE ENGINE ROOM. WE HAVE MANY PASSENGERS IN ALL ABOUT 90.

Distress rockets went up -- dozens of them -- but none were seen by other ships. Visibility was zero. Weak signals (obviously from a fading emergency battery) trickled on until 9 PM, then stopped. The search went on for seven more days, but nothing was found. We shall never know the agonizing details of the last hours of the *Hans Hedtoft*. We do know that Carl Johan Nilsen Dejligbjerg died at his post, trying to get help.

JOACHIM GEISSLER
1923-1963

JOACHIM GEISSLER (40) S.S. *Munchen*, June 25, 1963.

From this 941-ton German vessel Geissler sent out an SOS at 7:50 AM. Eight minutes later he was sending signals on the DF frequency of 410 khz and concluded with "It's very difficult for me to get to set." No further signals were heard, but that first signal had brought many would-be rescue vessels to the scene. They found 15 survivors; 27 were lost. The official inquiry said that a life raft was occupied by 11 members of the crew, including the master and a chief mate. They said that the radio officer (who had sent SOS signals to the very end) had lost his hold on his way from the radio shack and had fallen into the water. Attempts to rescue him by a fire hose were unsuccessful.

JOHN BRYAN HILLIAR
1944-1964

JOHN BRYAN HILLIAR (19) S.S. *Ambassador*, February 18, 1964. The British freighter was drifting helplessly in a storm-tossed Atlantic; engines disabled, forward hatches stove in and a 40-degree list. Her Sparks had sent an SOS in the dim light of early morning, and several ships had responded.

They fought through enormous seas to reach the Ambassador, but when they arrived the weather was so bad they couldn't launch boats.

The captain sent a final message, PUMPS CAN'T HANDLE WATER IN ENGINE ROOM ... CONDITION OF SEA VERY DANGEROUS ... HAVE NO LIGHT ... OIL FLOODING ENGINE ROOM.

In all 20 men were saved by rescue ships, but 14 others, including young John Hilliar, were never found.

A prolonged search of the windswept area got no result. It was John's first ship as a solo operator. Before that he had made one voyage of seven months as a second operator.

CARLOS RAMON GRIFFITH
1925-1967

CARLOS RAMON GRIFFITH (42) *Panoceanic Faith*, October 9, 1967. Carlos was born in Puerto Rico. When he sent out an SOS 870 miles southwest of Kodiak in the Pacific, the ship was taking on water in two holds, and being battered by 50-mph winds plus 20-foot waves. Five hours later, Griffith reported engine trouble, listing and sinking. A Russian, two Japanese, and a Coast Guard vessel were all on their way; three rescue planes, too. Five men survived (picked up by ships); 36 -- including Griffith -- perished.

JOSEPH VYHNAK
1928-1980

JOSEPH VYHNAK (52) S.S. *Poet*, October 24, 1980. Vyhnak had sent the routine departure message -- course, speed, destination, etc. Eleven days later, no more messages having been received, a search was started by 50 aircraft missions. A month later, the USCG said that all crew were presumed dead. Vyhnak was a native Czechoslovakian -- political prisoner in the USSR -- emigrated to the USA and became a U.S. citizen. He got his license, flew to Port Said to relieve a sick R/O, and began his ill-fated voyage from there. He'd overcome great obstacles to become a radio operator. The Battery Park Memorial is all that remains of him, his family, his whole story.

KARI BERGELIEN
1946-1973
S.S. Anita

BENTE KNUDSON
1949-1973
S.S. Norse Variant

March 22, 1973

Two Norwegian vessels sank about the same time in a storm off Newport News, both surrounded by a cloud of extraordinary coincidences.

Both had women operators, both had identical length, same cargo capacity, and both were carrying coal to Europe.

However, they were not of the same company. There was only one survivor from both ships: a motorman from the Norse Variant.

Both women operators had remained on duty right through the storm, though some of their crews had taken to the lifeboats.

They are heroines alongside their male colleagues.

DAVID STAIER (28) S.S. *Mezada*, March 8, 1981.

A native of Israel, David had been on several Israeli ships, accompanied by his wife Malka. (Israeli officers are allowed to take their wives aboard, and most do.) The ship was hit by a storm off the coast of Bermuda. Malka was saved, but her husband remained at his post as the ship sank and was never found.

DAVID STAIER
1953-1981

INTERNATIONAL MORSE CODE

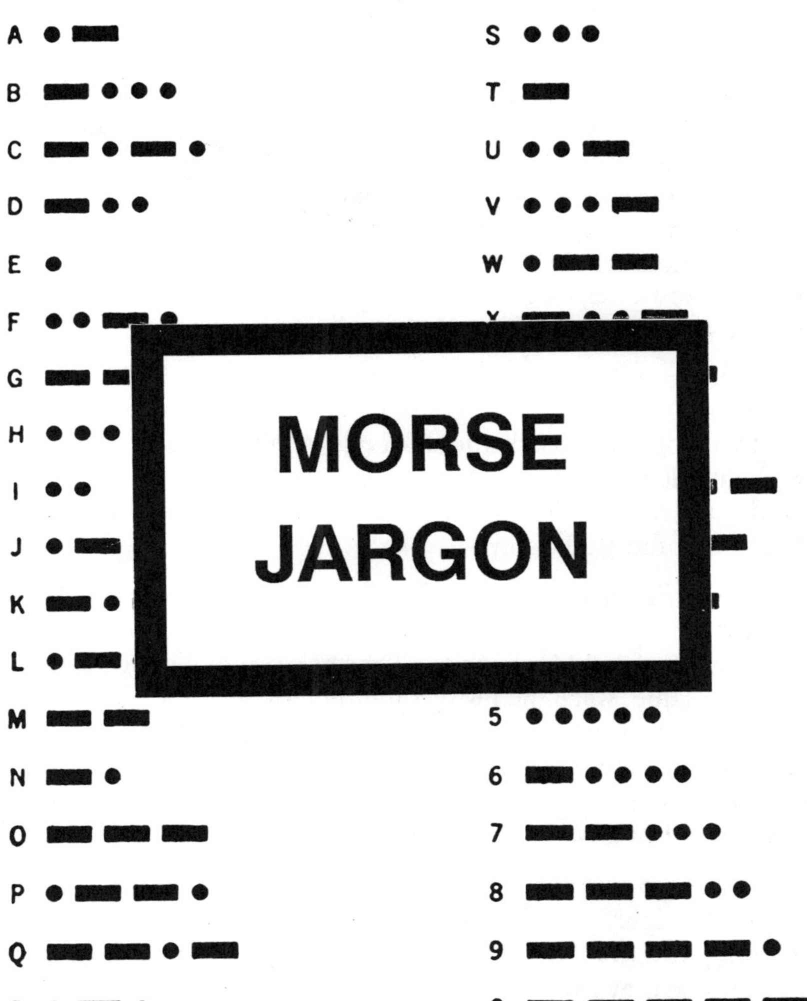

MORSE JARGON

Morse may be heading for extinction, too. It shouldn't. It's the most useful and understandable international language ever devised.

Scene: Anywhere at sea. Time: 1 AM

The second mate, scrutinizing a nearby ship, sees a light flashing from her bridge: "A--A--A--A--A." He scrambles for his blinker, gives answering flashes, then receives the usual query: "What ship?" In spite of being used to it, he often feels that this way of asking the question is, to say the least, rather blunt; in fact, rude.

His Sparks, making an encounter by radio, would not have been so awkward. Conversation would have begun with "GM OM PSE TR?," meaning "Good morning, old man. Please, what is the name of your vessel, where are you bound, and where from?" (The abbreviation TR is quite comprehensive.) This normal greeting on the international radio frequency of 500 khz has taken less time, but has requested much more information than the blinker signals did. Yet, though terse, it's polite and friendly.

Morse code is, as you probably know, the utterly simple use of short and long sounds to communicate on a radio frequency. Only two sounds are used: the short one is called a "dit" (dot) and the slightly longer one is called a "dah" (dash). With these two sounds in various combinations, letters, words and sentences can all be communicated easily.

The original version invented by Samuel F. B. Morse in 1835 was later modified to become the International Code of today, but often it's still called Morse code. The basic principle of organization is the same, since the most common letters (e ... t ... a ... o ... i ... etc.) have the shortest forms, while the uncommon letters (z ... q ... x ... etc.) have the longest. There are also dit-dah combinations for punctuation marks, numbers, abbreviations and even whole sentences. (Look back at the sentence in the second paragraph.)

The Q code is the official basis of marine communication. Q signals are highly specialized abbreviations of three letters, all beginning with the letter Q. Q signals mean the same thing in all languages, so they can be understood by any radio operator on any ship in the world. Here are some examples:

QSB Your signal is fading.

QRL I am busy.

QRP Decrease your power.

QTC I have (indicate number) messages for you.

Most of the Q code is suitable only for radio, but some signals would be welcome on the bridge. For instance, a slightly rattled new third mate might like to know and use "QRS" -- "Send more slowly!" Q signals are only the skeleton of the language in general use. English, as every sailor knows, is the international language at sea and in ports. Super-abbreviated English, as every Sparks knows, is the common language of all ships, whatever nationality, when on the air.

The answer to the request for a TR would include such time-savers as BND (bound for), FM (from) and end with civilities like GN (good night), or BV (bon voyage). No Sparks of any nationality, no matter how rudimentary his spoken English might be, would be puzzled by PSE (please), TKS (thanks), CFM (confirm), SIG (signature), or TRBLE (trouble). But perhaps "*no* Sparks..." is too all-embracing. First-trippers are inevitably mystified at first, and that goes for American, British and other English-speaking novices as well as those from other nations (who soon find they'll have to polish up their English). Their textbooks will have given them very few of these common abbreviations, and those probably in a shame-faced Appendix, like a dictionary's glossary of slang. A new man has to learn many things they didn't teach him at school, before he becomes a full-fledged radio officer.

Words are abbreviated in several ways. One way is merely to start the word and let the rest of it be understood. This is very useful when dealing with unwieldy polysyllables such as "transmitter," "press (news) bulletin" or "weather report"; they become, in turn, TX, PX and WX. Some words are only phonetic imitations of their originals -- CUD (could), WOT (what), U (you); a prize specimen of this variety is CUL (see you later). Other operators substitute letters with shorter Morse symbols, to save an odd second here and there: "for" becomes FER ("dit" (E) instead of "dah-dah-dah" (O); BI (by), MANI (many) and VERI (very) have changed "dah-dit-dah-dah" (Y) into the briefer "dit-dit" (I).

A common farewell signal is "73," meaning -- no one knows

why -- "best regards." It arose in the 1870s, when telegraphy was only possible by landline, usually between railway stations. It had -- it still has -- the comradely spirit of all Morse conversations. Radio operators everywhere, whether Sparkses, hams, post-office telegraphists or ultra-busy coast-station operators, do not in their successful efforts to clip words and save time let pleasantness be eclipsed by efficiency. Camaraderie is natural to people in the same trade, using the same language, tools, etc. Ships' operators have similar equipment, same problems -- and identical language. Typical calls to CQ (call ships) are: PSE ANIONE GOT TIMES PJC TFC LISTS?" (Please, does anyone have the times of PJC -- Curacao Radio -- traffic lists?" Traffic lists are sequences of ships' call-signs for whom there are messages.) PSE ANIONE GOT CFH WX? (CFH/Halifax weather report covers most of the Atlantic). PSE ANIONE QUA ......? (Anyone have info -- whereabouts, etc., of SS .....?). This preliminary PSE ANIONE .....? is in very many of the general calls to CQ on 500 khz, and usually someone comes up with advice. Hundreds of times daily, ships of different nationalities help one another in little ways, in a spirit the United Nations could envy.

International brotherhood -- altruism in any form -- has it ever been accomplished? Is it only a pie in the sky? The League of Nations was well-intentioned; the United Nations ditto, but the skyscraper on the East River holds more backbiting, distrust and narrow chauvinism than brotherly love.

The very first requirement for real understanding is a common language. Swift and accurate translations (as in the U.N.) do try -- but there's always that small time gap. At sea, the common language (for Russians, Australians, Greeks, Italians, whoever) is an abbreviated English plus the Q code.

Am I saying that international Morse code would make the perfect universal language? No, but here is an example that a universal language is not impossible.

The people who want to do away with Morse code and rely on those fragile computers will have their day, but I think that

Morse code still has some impressive features going for it: simplicity, reliability, directness. It's international and it gets through when other communications fail. Perhaps the most intriguing use of Morse code in recent years came out of the Vietnam War. Jeremiah A. Denton, Jr. was a prisoner of war being put through a sham television interview by his captors. He was supposed to agree with the interviewer that he was being treated well, fed and housed adequately. Well, that's what he said on camera, but he also blinked his eyes in Morse code, spelling out the word "torture." His message got through.

GM QRK? PSE OM HAVE U GOT NSS WX? (Good morning. How do you read me? Please, old man, have you got NSS -- Washington D.C. -- weather report?)

That was a *Dutch* ship asking a *Yugoslav* one for info!

When help is badly needed, when the signal SOS quiets all small talk on 500 khz, a Sparks's shipmates will agree with him that the flag of the ship in distress doesn't matter a damn. What's their position? Who is nearest to them? Can we help? Who will get there quickest? How long can they stay afloat? What is the weather like? When time is precious (especially if the distressed vessel cannot generate her own electrical power and is relying on fast-dying batteries for radio-communication), then the clipped Morse jargon is invaluable.

The most urgent signals, after SOS, are: XXX, for a man overboard or badly needed medical advice, some individual life in danger; and TTT -- navigational warnings such as floating logs, unlit buoys, etc. Whether the radio traffic is important or not, an easygoing atmosphere is always maintained. An amicable OM ("old man" -- *not* referring to the skipper or to actual age) is scattered throughout the exchange of signals, right up to the final, lighthearted "pom-tiddley-om-pom" from one key--which the other answers, of course, by "pom-pom."

You may now guess the reaction of a Sparks if he was greeted out of the blue by "What ship?" He'd feel as affronted as if he had been walking down the street and a stranger had rushed up to him, grabbed him by the lapels, and grunted, "Who are you?"

1840	Morse Code was officially born (patented).
1900-1912	CQD was the call for help.
1912-1999	SOS " " " " "
1999	SOS to be discontinued. (Morse to die?)

LONG HAIR AND TRASH IN SINGAPORE

You are the radio officer of a ship visiting Singapore. You're in the Hotel Raffles. You're sprawling in a lounge chair, under slowly revolving fans. As you sip your martini and look around at the grand furnishings, you have a dim awareness of all the famous people who have been in this famous place. You recall that Somerset Maugham said this hotel "stands for all the fables of the exotic East."

You're only 80 miles north of the Equator, but you're comfortable (temperature 79, humidity 60). The population of Singapore is two million (79% of Chinese origin). This independent island republic off the coast of Malaya is only 27 miles long and 14 miles wide.

"Another drink, sir?" The respectful, turbaned waiter goes off to get you one, and you continue musing about Singapore. Stamford Raffles founded it in 1819. It was a British colony until 1959 and has been completely independent since 1963. It was a very important British naval base, conquered and occupied by the Japanese in World War II. Now? Doing very nicely, thank you. As the world's fourth largest port -- huge shipping dues plus lots of tin and rubber exported -- it has few complaints.

Well, Singapore does have a couple of complaints, sort of. They appeared on your ship's notice board, a few days before you arrived. They came from Goh Choo Cheng, Port Captain of Singapore, addressed to "The Shipping Community." The first was:

> In line with the Government's decision that action should be taken with the LONG-HAIR PROBLEM, as part of a campaign against undesirable alien influences, the Port of Singapore would be most grateful for your cooperation in implementing this decision by advising ship's officers and crew coming ashore to be MORE TIDY IN THEIR APPEARANCE.

This sign is displayed in many Singapore shops.

OFFSET BY PRISON INDUSTRIES IN CHANGI PRISON, SINGAPORE

So the city fathers had decided that long hair (on men, of course) was the thin end of the wedge of "undesirable alien influences"? Perhaps they reasoned thus: "Hippies have long hair. Anyone with long hair is a hippie-suspect. Everyone knows about their anything-goes attitude. Their loose living goes by stages: from wild parties to marijuana to orgies to anarchy. It all *starts* with LONG HAIR. Let's stop the trouble at its source." The announcement on the ship's notice board had specified, exactly, what "long hair" was, and numerous posters made it clear.

(see illustration)

The immediate result of this notice had been increased business for the ship's amateur barber. The bos'n decided that his three-weeks-old beard would mark him as an undesirable alien, and ruefully reached for his razor.

When you came ashore you found that this long-hair veto was not restricted to seamen. Outside many shops was a notice: MEN WITH LONG HAIR WILL BE SERVED LAST! The definition and exact description of long hair (as given on the ship's board) were all there. The notice was repeated in each of Singapore's three languages: English, Chinese and Malay. A fourth language, line drawings (for the illiterate, or foreigners), made sure that no one escaped the message.

When you first saw these posters, you wanted to purchase some of them, to share your amusement with others. You peered closely at them, and found, in the bottom right-hand corner: "Printed and distributed by Singapore Prison and House of Correction."

Right here and now, sipping your martini in the Hotel Raffles, you decide to rent a car, drive to the prison, which turns out to be a long way out of town, pick up a load of these anti-long-hair things and do some sightseeing on the way. Wait, though. There's something else needs checking, first. It involves the second notice on the ship's board. This, too, comes from Port Captain Goh Choo Cheng:

SINGAPORE HAS ONE OF THE STRONGEST *ENFORCED* POLLUTION/LITTERING LAWS IN THE WORLD

Not one piece of trash or garbage is to be thrown over the side of the vessel in harbor.

The harbor is patrolled by boats using cameras with telescopic lenses to catch any violations.

Ashore -- do not throw so much as a gum wrapper -- if caught, you will pay a fine.

When you go ashore you will soon see: Singapore is a clean city without litter.

This last sentence needs checking out; you were so amused by the war against long hair that you forgot about the campaign against trash. You finish your drink, pay, tip and leave the hotel.

You are immediately beset by half a dozen drivers of "trishaws" or bicycle taxis. Perhaps it would be a good idea to trundle along in one of these things before trying to rent a car? It'll be fun. You'll not only see more, but you'll be more part of the scene than in a closed-in car, having to fuss about traffic. So you surrender to the nearest trishaw driver and get into the sidecar.

You roll along, past colorful Chinese restaurants, tall buildings, pretty girls dressed in both Chinese and European fashion, traffic cops, hawkers, etc. Well, well. The streets *are* clean. There just isn't the tiniest bit of litter, anywhere.

Singapore is a proud lady. She may have some funny ideas about long hair, but on the subject of trash, to quote Professor Higgins, "By George, she's got it." What? Make a fuss about one cigarette butt? *Yes*! Do the police pounce, give a ticket and fine there on the spot for one lousy gum wrapper? Yes! Yes! Yes! These are the thin end of a wedge which could turn Milady Singapore into as filthy a hag as New York City. (There can be no greater contrast. Even apart from her periodic garbagemen's strikes, New York is the world's dirtiest city, period.)

Let all the clean-minded citizens in all the cities of the world give a concerted cheer. Let the cheer roll in from all quarters of the compass, down the Malacca Straits and up from the China Sea, to swamp this little island with a tidal wave of enthusiastic approval. Hurrah for Singapore!

Singapore has had a glorious past, as previously noted, but one bad mistake was made. Prior to World War II, the British Navy positioned all the defense guns to point out to sea. They couldn't be swiveled around to face the rear; but, then, a land attack was not expected. In 1942 the invading Japanese army crossed the three-quarter-mile causeway joining the island to the mainland and took the city with very little trouble.

Singapore is taking no more chances: not with hair or trash, anyway.

P.S. You got those posters from Singapore Prison.

RADIO IN WORLD WAR II

Radio and its offshoot, radar, were vital elements on which victory or defeat hinged in World War II. Beside its obvious use for propaganda, radio (mainly via the BBC) was transmitting information to resistance fighters and spies behind enemy lines, as well as furnishing instructions and orders daily to governments, navies, armies and air forces. During the Battle of Britain, radio was an absolute necessity. Radar would put out the initial warnings of enemy aircraft approaching the coast. Then radio would alert the population (the howling sirens would be turned on) and provide information to RAF pilots and bases as the battle was joined in the air. This radio communication -- whether life-and-death matters or simple chit-chat -- was like a pulse of life, providing morale and news.

Radio was a lifelong interest of William Stephenson, the Canadian-born super-spy who was code-named "Intrepid" by Winston Churchill. Stephenson was the center of a revolving wheel of secret information throughout the war, when he commuted almost weekly between London and New York. His most important breakthrough -- not revealed, however, until the facts were declassified decades later -- was the development of a way to break the German codes devised on a complicated machine called ENIGMA. The Germans had constructed this code-machine so ingeniously and secretly that they put complete confidence in it. They used ENIGMA codes constantly throughout the war, believing that their messages could not be deciphered.

Actually, one of these coding machines had been captured and analyzed early in the war effort by the Allies. Recognizing that its value would last *only as long as the Germans didn't know the codes could be broken,* Stephenson clamped down absolutely on this information. Only a handful of people knew where his information was coming from. Most commanders in the field thought the information came from spies inside the enemy camp. Almost no one knew the German codes had been broken.

Part of the secrecy involved gathering a large group of translators, transcribers and various radio experts in a mid-England town called Bletchley. Here again, not even every person on the staff knew that this vital information was coming from the Germans themselves! Internal secrecy was stringent. The code-breaking *had* to be kept secret ... and it was.# The center of the activity was a large estate called Bletchley Park.

> The house in Bletchley Park was a red-brick Victorian monstrosity, just about the last place anyone would expect to find the keys to Hitler's day-by-day decisions and the enemy's inner secrets.
>
> Hidden in the rolling farmland all about were webs of radio aerials, already spread to net the faint murmurings of distant transmitters. At the center of each web were groups of experienced ships' radio operators. Stephenson, who had never forgotten his schoolboy exchanges with the Morse operators on Great Lakes freighters, regarded seagoing radiomen as among the world's best. They were accustomed to discomfort and to working in close quarters alone. They could hang onto the faint signals of a moving station surrounded by the clutter of other transmissions drifting across the wave bands. At sea, they had to recognize quickly the "fist" of the particular operator they might seek, detecting subtle characteristics in the way he worked his key that amounted to an individual signature. They had a sense of danger, important when later their transmissions guided the secret armies in Nazi Europe. Between the wars, thousands of such ships' operators were kept on a special reserve list in anticipation of a conflict fought in darkness.*

#*The British knew (from ENIGMA) that Coventry was to be a bombing target in November 1940; but they dared not evacuate the city and thus tell the Germans that their code had been broken. Coventry was sacrificed.*

**From* A Man Called Intrepid, *copyright 1978 by William Stevenson. Reprinted by permission of Harcourt, Brace, Jovanovich, Inc.*

The radiotelegraph operators who were part of the underground movement in Europe were called "pianists." Some were flown in from England. One of these was a beautiful young woman code-named "Madeleine." Her actual name was Noor Khan, raised in Russia with strong ties to France. Her father was from India, her mother was American.

The general opinion -- too late -- was that she should never have been sent on such a dangerous mission. Her haunting beauty put her in the "once seen, never forgotten" category. Still, she was determined to help the French Resistance. She was landed in France in July, 1943, and began her regular radio transmissions to Bletchley. In October she was captured by the Gestapo.

She had several sets of pills for emergencies. The "L" pills would kill her in seconds *if bitten* (just swallowed, they'd do no harm). These were reckoned to be the only real defense against a Gestapo interrogation. However, for whatever reason, Madeleine didn't or couldn't use them. She wouldn't give the Nazis any information. She was tortured and eventually shot at Dachau camp. Her heroism won her the highest award the British could give her -- a George Medal.

There were many heroic radio operators in WWII. It is fitting that Madeleine should represent them all.

GRAIN TO RUSSIA

For the U.S.S.R., it was the worst grain harvest in 100 years, but the U.S.A. had plenty. In Washington and Moscow a big deal was made. Twenty million dollars worth of wheat, corn, etc., was sold to the U.S.S.R. Ships galore (divided among the USA, the U.S.S.R. and other nations) would carry the stuff from Texas and Louisiana to Leningrad or to the Black Sea port of Odessa.

Steaming into Odessa came the SS *Manhattan*, largest ice-breaker in the world. She had cut her way through the Arctic Circle in 1969-70 to make the long-sought Northwest Passage from the Atlantic to the Pacific -- but that's another story. Now, in 1983, she brought 67,600 tons of grain -- much more than the average ship could carry -- to provide bread for the Russian people. Still this monstrous floating grain silo was only half full. She already lay so deep in the water (40 feet at the loadline) that most ports couldn't take her alongside. To have increased the load to her maximum capacity (114,000 tons, with a 50-foot loadline) would have made it impossible for her to dock.

The efficient Russian pilot came aboard. He was very polite and he spoke very good, efficient English. He steered the ship efficiently to an anchorage. Then left. Eventually (this was our first confrontation with Russian ideas about schedules and appointments) other officials boarded. "INFLOT" was the expressive name used by our hosts to refer to the strangers who "floated in." They were to take care of everything. They were the go-betweens for the port authorities, etc. They were very polite and spoke very good, efficient English. The net result, in the form of firm decisions and promises, was vague (*Mozhyeet-bit:* "maybe, perhaps").

The captain of the *Manhattan* had many questions. The Russians would rarely commit themselves. "Yes" might be dangerous. "No" could be impolite or discouraging. "Maybe" was just right. "Mozhyeet-bit" was to be the dominant word henceforth.

Q. When can the ship go into port and discharge there?

A. Very sorry, but ship too big. Maybe later...

Q. I meant that when we've discharged enough to lower the loadline, can we then go alongside?

A. Well ... we'll see. First things first, yes?

Q. How long do you think the whole business will take? Can you make a *rough* calculation?

A. Oh, very difficult to say.

Q. Longer than a month?

A. Mmmm -- well, maybe, maybe.

Q. In that case, is it clearly understood that there'll be a demurrage fee for every day over 30?

At this point, the clouds of uncertainty had crystallized into something positive. INFLOT, under the threat of having to pay a demurrage penalty fee, suddenly saw things very clearly. They started with discreet suggestions, then snowballed them into downright pressure on the captain to change the terms of the charter-party agreement. They wouldn't believe he had no power to do this. The captain hardened into simple negatives: "No. No. No. Again, no! *Nyet*. Nothing doing."

The *Manhattan* never did get alongside. She lay at anchor, slowly discharging her grain into barges by vacuvators (a type of large vacuum cleaner). Other ships came, docked, discharged and left. After nearly a month there were still 20,000 tons of grain left, with the loadline now at 30 feet. Considering this, plus the threat of demurrage, our hopes of getting alongside rose. They rose with (and like) the clouds of "maybe's" at every consultation with INFLOT. Then they evaporated. The *Manhattan* was part of the Odessa port landscape offshore for 45 days, a long time to be "on the hook" staring at Odessa a couple of miles away. You'd assume -- and you'd be right -- that officers and crew, having come some 5,000 miles to get under the Iron Curtain, would be curious to see things, to find out how the Russians live, to collect stories to tell back home. There was a launch service to the shore. It went back and forth five times daily, weather -- and the Russian sense of schedules -- permitting. But going to and coming from the shore could be summed up as a grim experience. Some didn't go at all. Some guys went once, period. Many -- over half the ship's complement of 62 -- spent a lot of time ashore. They had to be back on board every night because staying ashore was strictly forbidden. They'd forfeit their shore passes if they did.

Before we join this shore-hungry crowd ("What can we see? What's cooking? What can we buy? Where are the dames? Where's all the action, if any?"), whoa, there. No hurry. Gonna be here a long time.

After the VIPs, the only Russians who swarmed aboard were the stevedores. They were going to look after the mechanics of unloading the grain into the barges alongside the ship. As it turned out, they were also the representatives of the Black Market ashore. They paid well, in rubles, for chewing gum, blue jeans, panty hose and dollars, in that order of preference. The *official* rate for $1 was seven-tenths of a ruble, but in the Black Market that dollar would get 2 or even 3 rubles.

Item	Price, USA	BM Price, USSR	Markup
Dollar bill	$1.00	2 rub. ($2.50)	250%
Panty hose	$.50	5 rub. ($6.25)	1250%
Blue jeans	$10.00	60 rub. ($75.00)	750%
Chewing gum	$1.50	10 rub.($12.50)	1000%

These were prices on the ship, to the stevedores, with middleman profits. Ashore, the same items might get 50%, even 100% more.

Doing business with a stevedore was easy. Having attracted his attention, you'd point to some place on deck that was hidden from the soldiers who stood guard on the barges. Usually the soldiers were standing far below the main deck of the *Manhattan* and couldn't see a thing, anytime, but they might be high up on the bridge of a mobile barge, with a good view all round. Then, under some projection, in a deck house like the bosn's store or a paint locker, the Black Market flourished.

Official rules about money-changing were strict (allowed only in a bank, an INTOURIST office or a duty-free gift shop). But supervision was so low it was almost nonexistent. The authorities knew damned well what was going on. They secretly approved of this illegal importing of American goods, especially dollars. Anyone loaded with rubles would have to spend them. They couldn't be changed back into dollars without great loss ($1 would cost 3 to 5 rubles). So Russian things would be bought and used/consumed/exported. Everyone was happy.

Odessa is a drab city. The only relief to its grey overall appearance can be seen in the colorful pictures of Lenin. They are everywhere. Sometimes he is shown flanked by Karl Marx or various leaders; mostly it's just Lenin, Lenin, Lenin ... Taxis are communal things. You hail one. Unless it's full, the driver stops. You ask, or point to some paper on which your destination is written in Russian, if he's going your way. Then you'll either join the group or wait for another cab ... Restaurant tables might be shared, too, when the management is running out of space. Someone might come up to the table where you're sitting alone and rattle off some Russian that sounds like a leaky boiler (with all its zzzzzzzh's and sssssss's), while pointing to an empty chair. You smile and wave to show he's welcome ... There are two large department stores, each four stories high, where you can buy almost anything, but most of it is junk. A man's suit which would be overpriced at $10 in an American pawnshop (new, but shoddy cloth and crude tailoring) has a price tag of 155 rubles (over $200, official rate). Women's cosmetics, of poor quality, are very expensive.

Parades are paltry, amateurish affairs with kids and others marching to a small band with no uniforms. Even on May Day in Odessa, when Moscow puts on a big show, you see only a few flags and some extra pictures of Lenin. Several restaurants with dance bands are open until midnight. One is the recognized haunt for prostitutes. Seamen of all nations come there. The local "champagne" (not bad) is cheap and plentiful. Fights are inevitable. Secret police in plain clothes, but usually known to the local people, are everywhere. They seem to have the same outlook as the London police, who are content to have prostitution, the underworld and as much mayhem as possible confined to one area -- Soho. The idea is, since crime is inevitable, let's have it all, as far as possible, in one controlled area. The secret police also had a special reason for feeling they had everything under control; the town whores double as informers and spies.

Clean streets leave a nice impression. There's an official army of women, armed with whisk-brooms, brushes and pans, always tidying up; they are the Sanitation Department. Usually, though, the streets are very quiet. There is little traffic and very few private cars.

INTOURIST is the official travel agency. Its tour of Odessa cost 15 rubles, which was $20, official rate, but actually only $1.50 CGR (chewing-gum rate). In port I had met Ken Brown, an American civil engineer visiting Odessa, and we both wanted a tour. Provided with a lady guide, a taxi and driver, we set out.

Luda (35, married, mother of two, university grad, quite attractive) apologized for her quaint variety of English. We assured her it was charming. She admitted that Odessa was no great showplace ("Moscow, ah! And Leningrad, even better!"), but she rattled off her set spiels for various places in Odessa. Here was the famous Odessa Steps, on which citizens of Odessa were shot down by Cossacks as they went to greet the mutineers of the battleship *Potemkin*. This was in 1905, the first (halfhearted) attempt at a Russian revolution. Eisenstein's silent movie classic *Potemkin* (1923) was shot here on the Odessa Steps. At the top was the statue of the Duc de Richelieu, the man who rebuilt Odessa in 1820. Over there was the Institute for Eye Diseases, world famous for corneal transplants. That gun was a relic of the Crimean War. There was the Town Hall where delegates meet monthly.

But the pride of Odessa -- its biggest building, squatting on a hilltop like a mother hen -- was the Opera House. We *must*, Luda said, see the opera and ballet there. That grandiose structure with cupolas used to be a Russian Orthodox church until it was turned into an observatory. ("Is religion still permitted?" "Oh, yes, there are several Russian Orthodox churches, some Greek Orthodox ones and a small Roman Catholic one. And that," said Luda with smug finality, "is enough.") She gave us the official party line: with only old people attending the churches they'd soon die out, and religion with them. No need for persecution.

The next building was very busy on Saturday mornings. It was the Wedding Palace. We were lucky it was Saturday. Trickling out of it came an amusingly jazzed up version of Mendelssohn's

Wedding March, followed by a succession of brides, all dolled up in traditional white lace, carrying bouquets, etc. They were all going to the same place from here, Luda told us, and we should follow them. We did -- to the waterfront.

There was the Monument to the Unknown Sailor, killed in World War II. An Eternal Flame flickered beneath the pillar. Four kids, about 12 years old, two boys with unloaded rifles and two girls, stood guard. They wore a sort of Scout uniform. Four other kids were marching up, goose-step style, to relieve them. The guard was relieved every 15 minutes throughout the day from 9 to 5. All the assembly-line brides from the Wedding Palace came to leave wreaths.

Beyond it all, far out in the bay, we could see the *Manhattan* at anchor. I dashed back to the taxi to retrieve my Polaroid and nearly collided with a contingent from Cuba who had come to pay their respects. "Stand over there with Luda," said Ken. "I'll take the picture." I have it for the record: Odessa Bay, the *Manhattan*, the Monument to the Unknown Sailor, children guards, brides with wreaths, Luda and I -- all in a three-inch square.

Our little tour was ending.

"Well," said Luda, "what do you think of Russia?"

This was rather naive. How could we answer? Ken had flown over the steppes, which indicated the vastness of the biggest country in the world. I'd only seen parts of Odessa. We'd had almost no direct contact with the Russian people, to see how they ticked. Some living conditions had been rattled off by Luda as propaganda, e.g., cheapest rents in the world, 12 rubles a month for 3 rooms, free medical care, 24 days annual vacation for all, etc. But juggling with statistics and some sightseeing didn't get us closer to the ordinary Russian. I was in for a surprise contact.

TWO RUSSIANS SPEAK UP CLEARLY . . . PRIVATELY

I was walking along the street, carrying a long clothes bag. A young man and his girlfriend stopped me. They wanted to buy the bag! It had stamped me as a visitor, probably American. They wouldn't have risked being seen by secret police if we hadn't been the only people visible within a hundred yards.

I said I *might* let them have my clothes bag as part of another deal. I wanted some icons. These are religious pictures, usually on wood, dating from the Middle Ages up to Czarist times. Good ones are highly valued in the art world. Even mediocre ones could have antique value.

They had some, they said, back in her apartment.

"We get taxi. You no talk. We speak taxi driver. Careful -- you no talk!"

We stopped a cab. We shared the ride with others, in silence. The couple paid as we left. We quietly climbed up several flights of a building, passing many apartments, to hers. "Still not safe," they whispered; walls were thin. So we kept our voices very low.

Tretlana, 26, was a medical doctor now specializing as an optometrist. Sasha, 27, was a Jew hoping to emigrate to Israel now that restrictions were slackening. He had the icons for sale.

We bargained a while. He reckoned they were worth in the U.S.A. five times their cost. I couldn't decide whether their apparent antiquity would more than compensate for their poor condition. It was agreed that I should take them on deposit, see if I could successfully get them on board without detection, get other opinions on their value, then meet next day to settle the transaction.

The U.S.S.R. had forbidden export of icons. Their religious significance was considered irrelevant -- they were regarded as national treasures. If our business transaction with icons was discovered, Sasha might have been given five years in Siberia, and the least that could have happened to me would be to have the icons confiscated without compensation.

My immediate problem was going to be getting them past the soldiers on the liberty launch; then after them the ones at the gangway. But I hadn't much of a problem. The icons were small. I fixed them with masking tape into the lining of my disreputable (therefore Russian-looking and unsuspicious) gabardine overcoat. We went out for a meal.

After another silent cab ride, we snuggled into a quiet corner of a nearly deserted restaurant. We ordered a meal with a bottle of wine and at last relaxed. Tretlana and Sasha began to talk freely. How representative they were of the average Russian I couldn't know, but their pent-up feelings had found a safety valve.

They hated the system, they said. They resented their lack of freedom to move around to other towns and areas. They didn't believe what *Pravda*, which means "Truth" (!) told them. They couldn't do this, wouldn't dare do that. They took risks in the Black Market to get things which Americans took for granted. Fear of the secret police governed nearly everything, but there was no organized underground movement, only bitter grumbling.

Sasha said that at long last he had real hope of emigrating to Israel.

"But the officials -- the office people -- they take so-o-o-o l-o-n-g. The paperwork, regulations, the ... the ..."

"Red tape!" I interjected, to help him out.

Then I had to explain that "red tape" originally had nothing to do with communism. It was just the color of the stuff lawyers used, but now it happens to express the way Russian bureaucracy takes a year and a day to do the simplest thing.

I had my private opinion, partly based on Sasha (the man himself, not his opinions) about this sort of opening that the Kremlin was offering to Russian Jews. Here in the Ukraine, Jews were numerous. (Golda Meir had come from Kiev, the capital.) Who was being released by Moscow? Not the valued workers, but the "drones." Sasha, I fear, was one of them -- a nice guy with no real trade. I was to meet more would-be immigrant Jews and have this idea strengthened. Tretlana was a medical doctor; a highly valued person. She wanted to escape. But how?

"I have a Dream," she declared earnestly, "and a Plan to make it come true." (Her eyes, shining with hope, call for capitalization of Dream and Plan.)

"If I marry someone from an American ship, I'll pay him what he wants. Then, after a few months of 'red tape,' as you call it, I can get to the U.S.A. and we can get a divorce there."

Digressing from Tretlana momentarily, this marriage-of-convenience idea may have been the driving force for the three girls who were all set to marry three crew members of the *Manhattan* next time they came to Russia. A cadet on another ship couldn't wait; he got hitched to his lady-love in the Wedding Palace before he left. On the other hand, it might have been flaming passion or even love, the real thing. Whichever it was, the mechanics of marriage to a Russian woman were simple and cheap. A man and woman needed only 1 ruble 70 kopecks for registration, plus four letters or statements. These had to come from (a) the woman's parents/guardians if she wasn't of age, or someone in her family if she was; (b) ditto for the man; (c) the U.S. government; and (d) ditto from the Kremlin. All four letters had to be unanimous in their sanctioning of the marriage. The snag was that the Russian government wouldn't approve until the man paid for the *total cost of the woman's education.*

Back to Tretlana. The cost of *her* education, from kindergarten to university plus four years of medical college and two years at the eye clinic, would amount to a lump sum that would be enormous to a practicing Russian physician whose take-home pay is about 190 rubles a month. She evidently didn't know all this, or her Dream would have faded like morning mist. Nor did I, at that time. Hindsight wisdom has been applied to that feverish talk in the quiet little restaurant, with lovely, desperate Svetlana explaining that her only hope of escape was to be rescued by marriage to an American.

Was she indirectly suggesting that I might be her rescuer? Maybe (our old friend *"mozhyeet-bit"*). But this was only a passing thought as I explained *my* Plan. How, I argued, did folks such as Rudolf Nureyev, Stalin's daughter, athletes, politicians, pilots and other defectors do it? They were in groups of all kinds that went abroad. They were trusted. So -- how to get trusted?

"If you can't lick 'em, join 'em," I quoted. This cliche was new to them. It amused them. And there was truth in the jest -- only five percent of the Russian people are actually Party members.

There is another way for desperate Russians to escape. They stow away on foreign vessels and then plead for political asylum. The Odessa police know this. Obviously there are no statistics available, but this escape route is an alternative to getting over or under the Berlin Wall. That's why the docks were closely guarded. It explains the enormously bright searchlight traveling over the bay at night; this comes from a police patrol beat.

NIGHT ASHORE

The launch couldn't come alongside the gangway: water was too rough. No one could get back aboard. We returned to the shore. INFLOT wanted us to stay ashore for the night, but that meant they had to contact Moscow for permission. It was unusual, therefore almost unthinkable. The bureaucratic machinery took until nearly 3 AM to give the final "Da ... OK."

The hotel was old-fashioned, clean and fairly comfortable. I was allotted a big bedroom with the chief mate and second mate. Delightedly I found there was hot water for a bath, and I was soon warbling in the steam. There was a refrigerator that had nothing in it. The beds were comfortable enough, but the bed cover reminded me of Winston Churchill's summary of Russia: "It is a riddle wrapped in a mystery inside an enigma."

It had a diamond-shaped hole in the center, through which the sheets below could be seen. Apparently, whoever wanted to unmake the bed just put his/her hand through the hole, grabbed

the sheet, and pulled it out, leaving the top blanket undisturbed. However, to make the bed, you'd still have to strip it off the cover anyhow.

You figure it out . . .

THE WORST PART OF THE WHOLE VISIT

No doubt about this--the worst part of the visit was the damned launch service to and from the ship. Transferring from the rickety, swaying ship's gangway to the rising, falling, skeetering launch, timing a jump to meet outstretched helping hands, was unnerving enough. It was worse on the return journey, leaping *up*. Why nobody fell "in the drink" (no pun intended), especially when deadweight drunks were being manhandled onto the gangway, was yet another of those expected troubles that just didn't happen.

The hazards of joining/leaving the launch were only part of the miseries. To catch the last launch at 11 PM, you would have to start getting fidgety about 10 PM or earlier if you were in a part of the town remote from the docks. Then there'd always be somebody missing. The launch would wait; maybe 10 minutes, maybe an hour. The uncertainty was maddening. From beginning to end the 10-minute trip to the ship might be stretched into two hours or longer. At first, the guys who didn't get back in time did not have to forfeit their passes, as threatened; but later on, persistent offenders were confined to the ship. The nightly vigil of waiting for latecomers exasperated the "good guys."

When the weather was bad, the launch wouldn't come to the ship. If you were on the ship, you were stuck there; if ashore, you might have to be taken to the INFLOT-sponsored hotel for the night. You were legally ashore for that night.

The drunken, yelling, vomiting crew members on the last launch to the ship were something else. The launch generally was a grim experience. However, connected with it were always two soldiers assigned to the launch -- 18-year-olds who knew no English. They had to exchange "Z-cards" (Merchant Marine ID cards) for Russian passes, and vice versa on return. This was a simple job requiring only a few seconds, but they took their time. They'd look hard at the face of the guy before them. Then they'd slowly, carefully compare him with his photo on the pass, several times, back and forth. Eventually they would surrender the pass to him,

(Hidden on my back is an ikon. Soldier suspects nil.)

but very slowly, as though they were toying with the idea of having yet another checkup. Instead of saving time on the return journey from shore by doing all this en route, they'd wait until the launch was heaving up and down by the ship's side, with everyone lined up and impatient, to begin their slow, careful checking.

Their checking of other things was confined to things taken ashore. Apparently their orders did not include checking of items bought from shore, so dozens of icons, some carelessly wrapped and almost visible, were taken aboard. They didn't seem to know that icons were forbidden export--or did they? The authorities may have known quite well what was going on. With the trade agreement in the offing and goodwill to be shown, they may have told the soldiers to ignore things. My nervous devices (putting icons into coat linings or having them under my shirt, pasted to my body) were maybe unnecessary.

"P-P-S-S-S-S-T-T!"

This hiss to get your attention--it could be on a dockside anywhere in the world. You look around, to see the hisser; he jerks his head back, to summon you to him in his dark corner. He then swivels his head around, to check that no one is looking. You approach him. So far it has been standard Black Market routine. But now comes something different.

"You got dollars?"

The devalued dollar is a poor thing compared with the mark, the yen or gold. Most places nobody wants it, but they do in Russia.

PHOTOGRAPHY IS SUSPECT

A street photographer took a photo of me outside the Opera House. I paid him. By his sign language I gathered that the finished product would be sent to me by mail, so I was to address the envelope he provided. I printed my name and ship c/o INFLOT, Odessa. I should have gotten it back in five days, according to his five extended fingers. Uh-uh. Ten, twenty days later I still hadn't got it.

I used to rage at the photographer whenever I passed him, but he'd spread his arms in a gesture of innocence and made signs of posting-into-a-mailbox. Not having a receipt for the five rubles I'd paid him, I decided with a shrug that I'd been a sucker.

Thirty days passed. I'd stopped hoping. Forty days. I was resigned to the loss of my Rs 5. The day before sailing, my self-addressed envelope was delivered to the ship by INFLOT. I reckon this is what happened:

I'd heard that although mail coming into the country is censored in Moscow, local mail is dealt with locally. I should have received it in five days or so. But my foreign (Roman letters) BLOCK CAPITALS had made the letter suspicious.

The photo postcard of me grinning outside the Opera House (with a huge picture of Lenin looking sternly over my head) must have been to Moscow and back. The censors were in no hurry to establish its innocence.

NEWSPAPERS

The only English language newspapers available are the London and New York editions of the *Daily Worker*. Dictatorships never take the risk of adverse criticism. The ordinary Russian sees nothing but Pravda. If he has a local newspaper it only echoes Pravda. Privately he mocks it. He is news-hungry.

WOMEN IN USSR

Women's lib shouldn't have any complaints here. A female finger is in most pies, though women seem scarce as VIPs of the Kremlin and totally absent as international bigwigs. As street cleaners with whisks and brooms, as uniformed traffic cops, or as doctors (90% of the MDs are women), they're more than equal. They're ahead. Their retirement age is 55 to men's 60.

Waiting for An IKON Contact

Right Way
Quiet
Unobtrusive
Russian-looking

Wrong Way
Pipe
Attitude
American-looking

**Ikoneering has its own rules.
The most important is to be
UNOBTRUSIVE.**

Precautions for secrecy ruled the search for, and bargaining over, icons. Conversations in public had to be brief, whispered. Even at a private rendezvous we had to be alert for people who might be the secret police. Above all, I should try not to wear, carry or do anything that would stamp me as an American.

Once, while waiting for someone on a park bench, I was sprawled out, feet propped up on the armrest, arms dangling, very relaxed. My contact, when he arrived, was horrified. That any-old-how posture, that don't-give-a-damn attitude was very American, most un-Russian and a dead giveaway.

SAMOVARS

These are charcoal-heated metal teapots, huge things several feet high. They're very desirable as antiques and they're not yet on the Kremlin's forbidden list. Frankly, they are ugly, but fascinatingly ornamental in the Victorian style. They might be made of silver, they might be 80-100 years old, they might have various exhibition stamps on them -- for any of these reasons their rarity and value increase. They're hard to find for sale, and many are still in use in cafes and hotels.

RUSSIAN TALK

There are 35 letters in the Russian (Cyrillic) alphabet. Many of them look Roman but they're not the same. Their C is our S and their P is our R; hence CCCP on the side of Russian spacecraft is really SSSR, their equivalent of U.S.S.R.

The dominant sound in Russian talk is ZZZZZZZH or SSCHHHHH or just ZZZZZZZZ. ("BolSHoi" -- big. "NaZZ-Derovia" -- good health. "BoYSHoya-SpaSSebo" -- thanks a lot; and "MoZZyeet-bit" maybe -- are all examples of this strong sibilance.)

Apart from this hissing and buzzing, the inflection of Russian is similar to English. They pronounce English words and phrases correctly, with little accent. We have no trouble pronouncing Russian words; whereas French/Italian/Spanish (musical stuff through the nose or back of the palate) or harshly guttural German -- are not so easy to us.

THE ONE GLAMOROUS SPOT

As mentioned, the pride of Odessa is its Opera House and what goes on there. The ballet is first class. Orchestra, ballerinas, timing, acrobatics, scenery, dresses, lighting -- all are breathtaking. The interior is unusual. Instead of a mezzanine and lower balcony, there are two complete tiers of small boxes, each holding about 10 people. They are carryovers from Czarist days, to seat the upper classes. These cost Rs 2.50, but the orchestra stalls have a better view and are cheaper. They're right up front and cost Rs 1.80 (only two bucks and change, officially, or 30 cents CGR!).* From 7:30 PM to 10:15 PM each night the house is packed. Most of the ordinary Russian's basic needs have already been met, but the panorama of his life is like a canvas of grays and semitones. Across this canvas the ballet at the Opera House paints a broad streak of exciting colors and shapes.

TRAFFIC LIGHTS

The timing of traffic lights in Odessa is all wrong. At each changeover, after simultaneous yellows, the reds and greens also change simultaneously instead of being staggered. Greens should, of course, be withheld from stopped traffic as long as possible; if this wasn't done in the U.S.A. there would be a fantastic number of intersection accidents, chiefly involving those who race to beat the lights, or crossers-on-the-yellow. But not so in Odessa. They get away with unstaggered lights because there's very little traffic, the citizens are very law-abiding, and no one's in a hurry.

---

*Chewing-gum rate

ORDINARY FOLK

One scene I'll never forget. People liked to walk on the promenade near the Odessa Steps in the evening. Somehow I got talking to a man who was keen on improving his English. We were soon surrounded by about 20 others who were hanging on every word and raising their own questions about the U.S.A.

There were some naval cadets who had been to the Merchant Marine Academy in Kings Point, Long Island, and had visited New York and Baltimore.

The subject of Alaska was amusing. It used to be called Seward's Folly when he bought it from Russia in 1867 for $7,200,000. Pravda was now apparently suggesting to its readers that it had really only been leased. Therefore it was stolen! The atmosphere was friendly. These intelligent people were rebelling against the crude propaganda fed to them.

Topics like cost of living, crime, traffic deaths, Watergate--all derogatory to the U.S.A. and plugged by Pravda; these people wanted firsthand info.

There were no beggars. No poverty was visible anywhere. There were inevitable drunken fights in bars and restaurants, and occasionally seamen loaded with drink and money were mugged. Otherwise, there were no ugly incidents.

Odessa folk everywhere were pleasantly courteous. This was particularly so with the polite personnel of INFLOT. Pressured as they were from above, by port authorities, the police and ultimately, of course, by Moscow, they did their best to keep everybody happy.

Let's get the balance straight. On the whole, the visit to Odessa was quite pleasant. Furtive manipulations to accumulate icons were not the keynote. Fear of the secret police, the penalties for this or that--all came to nothing. There was no real trouble, no friction. With imagination working on uncertainty, it was the idea of trouble, the threat of "if", "when," etc. Let's liken this fear of grim possibilities to a black cloud in the sky which never actually develops into rain. The actual weather was like that, too. It went from neutral to pleasant. When the *Manhattan* arrived in Odessa, the trees were bare. Spring comes late there. At departure, 45 days later, the sun was in full command, with leaves and flowers everywhere.

There was no anti-Americanism. People were as pleasant and curious as they dared to be in public; they opened up in private. In restaurants, people sitting nearby would try to bridge the language barrier with smiles, gestures, toasts, sketches on paper. If they had a few words of English or were studying it (many were), they were glad to get practice.

GOODBYE, RUSSIA

Going back to the theme of expected-troubles-which-never-happen, the biggest anticlimax came at the very end, when the ship was due to sail. When the ship first arrived, everyone had submitted a complete list of personal belongings which had any resale value, such as tape recorders, cameras and money of any kind. We'd been warned there would be a complete check on everything declared when the ship was ready to leave. Everything would have to be accounted for. We'd have to have receipts for dollars changed at a bank or duty-free store. Severe fines would follow if everything was not just right. Other ships had been through this mill. It had taken as long as eight hours for a ship to be cleared, after thousands of dollars of fines. That's what we'd heard, anyway.

Well, the officers and crew of the *Manhattan* had had a good time, spent all their money, and it seemed doubtful if a single one of them had a receipt for anything. Thorough searching of the ship was expected. (The radio officer had many icons hidden in various places.) There was really going to be hell to pay, if...when....

What actually happened? Absolutely nothing. The ship was "cleared" by the stroke of a pen. There were no inspections, no checking. Nothing. Why?

The captain reckoned that his soft-soaping of the customs folks plus great generosity with cigarettes and booze to all VIPs had done the trick. Maybe they helped, but a month later news about a new U.S.A./ U.S.S.R. trade agreement -- details of which were being ironed out just as the *Manhattan* prepared to leave -- threw fresh light on that incredibly smooth departure of the ship. With discussions in a touchy stage, it wouldn't have helped to "rock the boat" by trouble over an American ship, would it? (The deal was an exchange of billions of tons of Russian natural gas for various U.S. products--which didn't include chewing gum and blue jeans.)

At this time preparations were also being made for Comrade Brezhnev's visit to the U.S.A. To cushion his way there, to oil diplomatic gears, a blind eye may have been turned on the *Manhattan* and the doings of its crew. But looking for the why-and-wherefore is speculative. What actually happened? Who knows?

QTO ("I'M LEAVING PORT")

In Odessa it was the middle of the night. What little night life there was had ended. Out in the bay the police boat roamed with its massive searchlight. On the *Manhattan* only the second mate and a couple of sailors were on watch, on deck; down below a skeleton crew of the black gang kept the engines turning. Everyone else aboard was asleep. It was 3 AM. Suddenly a boat crept out to the ship from the shore. News! We're cleared! Let's go! Haul in the anchor! And SS *Manhattan*, empty of cargo but full of memories of Russia, and with lots of Russian mementos, stole away.

CAMEOS

These "cameos" are sort of snapshot memory-pictures that are condensations of certain periods in my life. They're the essence of a certain time/place/mood/outlook.

Scene is the Lyons Corner House all-night restaurant in Coventry Street, near Piccadilly Circus. London time and date: 2 AM, February 7, 1949. I had missed the last underground train to Swiss Cottage (London area). (They stop at 11 PM for checkups, cleaning, etc.) I was dissatisfied with my job in the advertising agency of Lever Brothers. Unilever House is an imposing structure near Blackfriars Bridge, overlooking the Thames, but I felt shut-in, restricted to pettiness everywhere. I wanted *out* -- somewhere, anywhere. Now here I was, talking to a Canadian radio operator from a ship that went all over the world. That was it! Why not? I already knew the Morse code. I could study technicalities, get a license, escape! . . .

Nine months later, I'm in a smelly part of London's East End (gasworks nearby). I'm in a local government job, doing boring paperwork. The little memory picture is of me walking away from the office, after another pointless day. By this time I'd passed exams, etc., and had my license. I was ready, but the Marconi Company hadn't got an opening for me yet. Oh, to be in an interesting foreign port, going down the gangway, away from the ship where I was boss in my department, and responsible only to the captain -- instead of *this* (contemptuous glance around).

Three months later -- I'm in another job; employed by the Fulham Power Plant, on the banks of the River Thames -- with ships actually coming to a sort of dock at one end of the building. I'm a nobody, an office boy who collects outgoing memos and mail, waits to be summoned for errands. I am still waiting for Marconi's office to call me. In the meantime, I can saunter onto the ships that come right up to our back door -- see the radio gear -- talk to the radio ops--imagine that I belong there, instead of being stuck in a circle, the center of which is my job. Someday, that seat before the radio console, those knobs, those earphones will be the center!

It is 1973. I'm looking out of the radio-shack porthole at Odessa Harbor. This Russian Black Sea port is full of ships, and they are full of grain. They are waiting to go alongside and discharge it. This will be done by having it sucked out of their holds by "evacuator" machines. Black market trade with the stevedores will get 10 rubles ($13.00) for a carton of chewing gum. One hundred percent profit. Things ashore are either very shoddy or fantastically expensive or unobtainable.

It is 1954. I'm looking out of the porthole at Rotterdam Dry Docks. Everywhere is the Dutch obsession with work -- noise with drills or cranes. I'm busy dressing myself to go meet my Dutch sweetheart. The tape recorder I bought in New York is playing Strauss waltzes that will be forever entangled with this scene. Nelly is free for the weekend. So am I -- the ship won't sail before Monday. We'll take the train, as usual, to Flushing. (The captain decided to sail on Sunday. When I came back, the ship had gone! They bundled my things into a sack, grabbed another radio operator and left.)

It is 1967. Out there, through the porthole, I see St. Nazaire, France. On the main street is the car-rental place I'm shortly heading for. I'm going to drive to that little village where M'sieur and M'dame, in their little hotel, will soon, smiling, present me with a delicious meal. I'll wash it down with their excellent table wine, try to converse using my rudimentary French plus gestures plus pictogram-sketches on paper. Then I'll discreetly collapse, belly full, into slumber. No one is in a hurry in this Brittany farmland. It has been that way for centuries.

THE KING OF THE RADIO SHACK

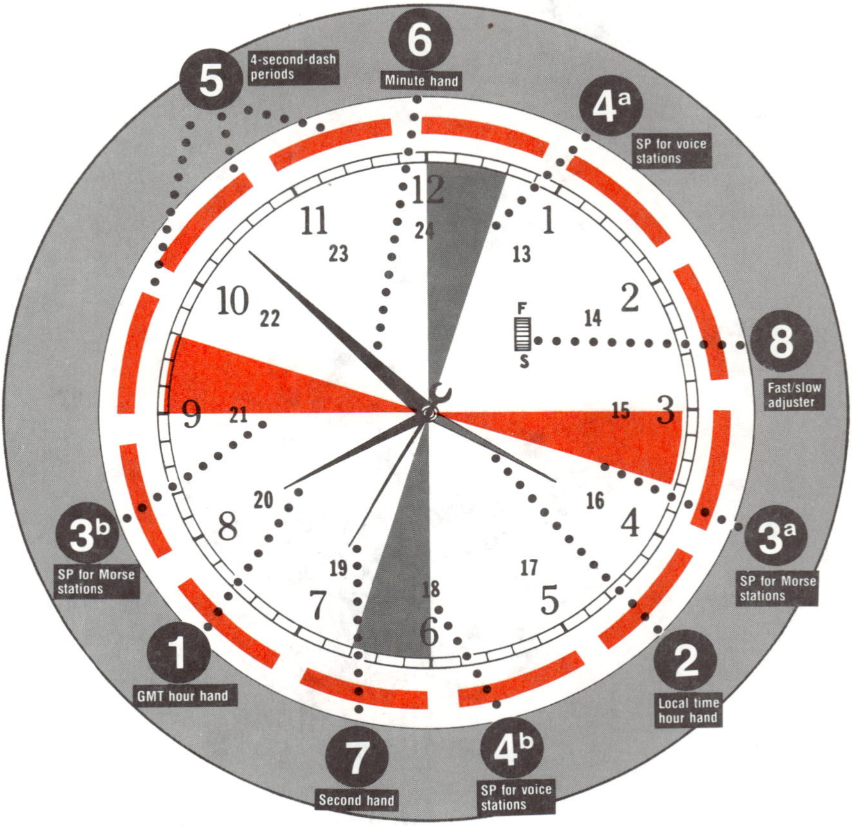

Explanatory notes

1. GMT (Greenwich Mean Time) or UT (Universal Time) or Zulu Time (1630z = 1630 gmt) are abbreviations of what is accepted internationally as **THE standard reference-point** — along with the Greenwich Meridian (North Pole — London — South Pole). It was that way for many years before it became official in 1884. Navigation — star-tables — astronomy — RADIO — all need it and use it constantly.
2. Local time changes 1 hour every 15 degrees longitude. This hand swivels independently.

3a. The Silence Period (SP), listening for ships in distress, might almost be called The Holy Period. Woe betide anyone who accidentally or carelessly touches the key during this time. SPSP's (silent-period special-policemen) pounce on the offender with QRT! (stop sending) or just SP!

4a. This period isn't nearly so sacrosanct. Many ships don't have it marked at all.

5. To activate auto-alarms on all ships within about 1000 miles, four-second dashes followed by one second of silence are sent. This sequence repeated three or four times will usually do the trick; but to make sure, twelve consecutive daaaaaaaaahs (4 secs.) are sent. The red bars are watched to help the timing.

8. Accuracy of time-signals (sent from places all over the world) vary from a tenth to a millionth of a second. Super-accuracy isn't required, but daily checking is desirable. The different techniques used by different countries vary interestingly. CHU/Montreal is introduced by voices speaking French and English. As is fairly well known, the French/English situation in Quebec can be uptight;* so to prevent either side feeling aggrieved, the sequence goes French-English then English-French for every two minutes right through the day. *("Why should we be Number Two?")

TITANIC

The Edwardian Era which began the 20th century was peaceful, humdrum. Across this tapestry of gentle pastel colors was slashed a vandal's brushstroke of violent red -- the *Titanic* disaster. At 11:40 PM on Sunday, April 14, 1912, an iceberg scratched the side of the ship on her maiden voyage. In two hours and 40 minutes, she was sinking two miles down to the bed of the Atlantic, and 1,517 people were dead or dying.

The basic framework of the calamity is:

April 14	11:40 PM	iceberg collision
April 15	12:15 AM	first SOS
"	12:45 AM	first rocket
"	1:10 AM	first boat away
"	2:05 AM	last boat away
"	2:17 AM	*Titanic* upends
"	2:20 AM	*Titanic* sinks

The tale of the *Titanic* -- almost the mere word -- fascinated four generations through 73 years. Then -- SENSATION! On September 1, 1985, she was found, on the seabed, much of her intact, with everyday things like plates, corked-up bottles of wine, chamber pots--even an ordinary teacup that had fluttered down two miles and was sitting primly on a huge boiler. They were not just described, they were photographed. An old, old story had become here-and-now, front page, topic of conversation.

There were some new facts -- not many -- to be added to what was known: she had broken up as she went down, so fore and aft sections are separated by six hundred yards; there's no gash on her right side, so the iceberg didn't rip her open -- it just popped all the rivets to let the sea flood in. There are no bones of human beings to be found -- mollusks and deep sea corrosion have finished them completely. With so much that is old and so little new, what hope is there of reviving or exciting interest? This account is going to try in three ways.

(1) Radio, or "wireless" as it was then called, is well known to have been an important factor. There were four wireless operators directly involved: the *Titanic* pair (Phillips died, Bride

The "unsinkable ship" leaving Southampton, April 10, 1912, bound for Cherbourg (France) Queenstown (Ireland) and then to 41.46° North 50.14° West at which place she sank, at 2.20 am April 15th 1912, and was next seen (via TV cameras on Dr. Robert Ballard's "Argo") at 0048 am Sept 1st 1985. A gap of 73 years 4 months 16 days 22 hours 18 minutes.

Radio Gear
(most powerful of any ship then afloat)
5kw motor-generator 300v 60 cycles from 110v DC or 24v batteries
Transmitter 10" induction coil
Receiver Marconi magnetic detector + multiple tuner (100-2500 metres)
Marconi rotary discharger (to eliminate spark-gap damping)
Range 250-400 miles (day), 2000 miles (night)
Antennas 200 ft high, 600 ft apart
Twin "T" (transmit/receive) Insulated cable to ground (hull).

Technical Details of Ship
length 882.5 ft
tonnage 66,000 (displacement)
propellers 3
passengers 2340
lifeboats 20
(rear funnel a dummy, used for storage)

survived); the *Carpathia* operator, Cottam (he accidentally picked up the SOS, and thus saved 712); and Evans of the *Californian* (who'd gone to sleep just before the first SOS and wasn't awakened until it was too late). This time the tale will be told from the professional radioman's viewpoint. What about those three iceberg messages that never got to the bridge of the *Titanic*? Were they vital? (Or, conversely, was their absence deadly?)

(2) Captain Lord of the *Californian* has, to most people, been the villain of the drama who could have rescued all if he'd responded how and when he should have. The theory you're about to read is not quite new, but it's had very little airing. Lord's true character, intentions and actions are all quite irrelevant. You can think what you like about him. He COULDN'T have saved those doomed to die on the *Titanic*.

(3) There will be a different method of presentation. The account of the two hours and 40 minutes that the *Titanic* took to sink--plus the preceding day, from 9 AM on--will all be in present tense. We're going to LIVE each passing moment, knowing nothing of the future. There will be comments, historical perspective, etc., but they'll be in italics. Both accounts -- the here-and-now version and the lofty, God's-eye view -- will progress remorselessly along in a single time frame. Instead of separate accounts, you'll have the overall picture of what was happening on each ship, minute by minute. (Small clock-faces will keep track.)

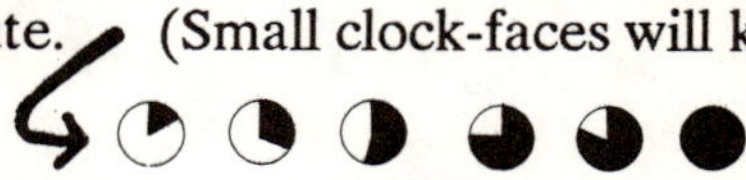

What should have been done -- or not done -- to prevent this most famous and most spectacular of all sea disasters? Twelve "IF..." clauses will show what might have changed everything and greatly reduced or eliminated the death toll. Blame for the disaster is not going to be pinned on anyone or anything! Circumstances, happenstance, the pressure of doing what is normal, to be expected -- yes, they can be determined.

Having established the general area for blame, though, we'll find there is no one person who can be singled out, accused and found indisputably guilty. All this must sound like a contradictory plea of "guilty but innocent." To clarify this, we'll deal with the very first accusation, the most obvious and simplest: there were not enough lifeboats for all on board.

Let's go back to a dinner party in London in 1909, where the idea of the *Titanic* and her sister ship, the *Olympic*, was born. Some very interesting talk at the party and what happened thereafter were dug out of musty files and displayed on British and American TV in 1983.

The most important people there were White Star boss Bruce Ismay and Alexander Carlisle, shipbuilder of Harland and Wolfe's Shipyard, Belfast. They were planning the two most luxurious floating palaces in the world. There were going to be swimming pools (first ever on ships), a French restaurant, Turkish baths, squash courts, etc. However, there was lifeboat accommodation for only 53% of those aboard. The ships, certified to carry 3,547, had boats for only 1,178. When the disaster did occur,

IF (1) **there had been enough boats and rafts for all aboard, the slow rate of sinking plus perfect weather would have enabled all to be saved.**

This lack had irked Carlisle at the dinner party. The Board of Trade rules (governing British ships) were absurd. Irrelevantly connecting the number of lifeboats needed to the gross tonnage of the ship, they required the *Titanic* to have only 16 lifeboats. Carlisle was very interested in revolutionary new davits made by Axel Wellin in Sweden. Each would hold up to four lifeboats at once. Sixteen of these, argued Carlisle, would give the *Titanic* up to 64 boats each -- safety for all passengers and crew. Also, forestalling the inevitable change in those B.O.T. laws of 1894, they'd be money-saving. (*That* was the bait, then/now/always.)

Bruce Ismay was managing director of the White Star Line and chairman of International Mercantile Marine (I.M.M.), the American conglomerate backed by Pierpoint Morgan. As the legal owner of the *Titanic*, he agreed with Carlisle's request for the new davits, but balked at providing more lifeboats. Carlisle wanted 48.

At the British Inquiry after the *Titanic* had foundered, Carlisle said he'd tried and tried to get 48 lifeboats on her. Ismay denied knowing anything about this.

Seventy-one years later (1983), the BBC-TV commentary "The *Titanic* -- A Question of Murder" uncovered his lies. "Not enough lifeboats" was given as a basic cause for the huge loss of life, but lies and coverups had prevented anyone from realizing that there could have been enough. If a few more dollars had been invested, everyone could have been saved. However, Carlisle was not the possible hero who could have saved all (as the BBC implied). He went part of the way, then chickened out. To vary the metaphor, he was a roaring lion in his own sphere (the shipyard), but a meek kitten when dealing with his client, Ismay.

So who can we blame? Carlisle, for not pushing hard enough? Hardly fair to the man who knew what should have been done but hadn't enough spunk in his character to finish what he started. Even then, he'd have had to out-argue Ismay. Shall we blame Ismay himself? Or the Board of Trade, for having such obsolete rulings about lifeboat requirements? When we try to fix the blame, we just can't. How about the White Star VIPs (of whom Ismay was the most important)? They were actually going beyond the inadequate B.O.T. rulings! The *Titanic* had room for 216 *more* than the laws required. With an eye to the future, they were talking to Carlisle, and considering his revolutionary ideas. How could they be the Bad Guys with these good ideas? The *Titanic* was called "unsinkable" anyway! So judge the company (Ismay particularly) by those positive actions and those good intentions, with their apparent relation to the future at the time--not by hindsight wisdom, or swooping down on them after the disaster.

The 1894 rule "for vessels of 10,000 tons & upwards" was being obeyed. (Other ships were similarly at fault. The German *Amerika* could only accommodate 55% of those aboard; Cunarder *Carmania* could take only 29%.)

It's the old, old story. You have a complaint, so you try to fix the blame on someone or something -- and you end up fighting an abstraction. This is certainly applicable to the *Titanic* in 1912. Things that were definitely wrong couldn't be pinned on anyone.

The next IF is also a glaringly obvious mistake, and just as difficult to pinpoint on any person --

IF ② **the *Titanic* hadn't been tearing along at 24 knots (that's over 27 miles per hour) into and through very dangerous ice fields and iceberg areas, barely acknowledging warnings....**

Without pressures of schedules, commitments, expectations of what was to be expected of the largest, most luxurious, maybe fastest liner in the world on her maiden voyage, Captain E. J. Smith, commodore of the White Star Line, might have done what Captain Lord (*Californian*) did: take sensible precautions and anchor for the night. He'd been warned six times (though he'd only received three of the warnings). Any captain in his position today would be considered criminal to have steamed along at that speed in an iceberg region.

Speed is an easy scapegoat, but veteran captains then and now swear that it always will be their universal practice to maintain course and speed as long as the visibility is good. Schedules have to be maintained. Business deals might depend on time of arrival, and passengers are always in a hurry to reach land. The perennial philosophy of captains and companies is "one can't slow down or change course because there may be ice somewhere ahead."

Now that we've brought Captain Smith into the picture, we're reminded that the difficulty of fixing blame doesn't apply to maritime situations. The Master of a ship is both the hero and the scapegoat; he takes blame/credit for anything/everything. So, we can zero in on Captain Smith. But can we? Should we? Postcards printed just after the disaster showed Captain Smith and wireless operator Phillips as two heroes of the *Titanic*. They were portrayed almost as saints. Yes, they had gone down with the ship, but suppose Smith had been picked out of the water? Suppose his attempt at self-sacrifice had failed? Suppose he'd had the gall, the indecency, to be still alive when 1,517 of his passengers and crew were not? There'd have been questions to him in the witness box about lack of caution. The Master of a ship is responsible for everything. Ergo, Smith was responsible for 1,517 deaths? This is an oversimplification, and it just won't wash. Let's salute Captain Smith, then continue.

I've never been in an accident —
Never saw a wreck —
Never have been wrecked.

Captain E.J. Smith's answer to question about his 40 year career at sea in 1909. (He was 56 then.)

He was "fearless."

That was the trouble.

His attitude infected his navigating officers.

If he'd been like Captain Lord, prudent, cautious, he'd have taken alarm, and anchored for the night.

However, he had the decency to go down with his ship and 1,517 passengers/crew.

But suppose he'd been picked out of the water, survived, and had been at either or both court hearings, USA/UK?

IF ③ **a number of attitudes had been different, the chances of the accident happening would have been slight.**

Captain Smith was a focal point, but we can't blame him personally. Most captains were like him. This one, cold night in 1912 would change everybody.

Here we can include two false IFs that concern what were considered faults in the ship's construction.

> . . . **if** the *Titanic* had had the "inner skin" other liners had. . . .
>
> . . . **if** one transverse bulkhead had not stopped short of going on to "D" deck. . . .

These old theories are now discarded. The iceberg's wound let the water come UP over and past the waterproof compartments, and an inner skin wouldn't have helped, either. They're mentioned here only because they were favorites for such a long time.

Two more IFs that *could* have stopped the tragedy were linked to Frederick Fleet, the lookout who first spotted the iceberg from the ship.

IF ④ **he had sighted the iceberg a few seconds earlier, the collision could have been averted entirely by a simple maneuver.**

From the sighting to the moment of collision, only 37 seconds elapsed. The quartermaster at the wheel swung hard left (obeying the yell of the chief mate), but this simply turned the liner far enough for it to experience a glancing, 300-foot-long collision just below her water line.

Fleet survived the sinking. At the American inquiry, he said he had asked for binoculars, but was told there were none available for the lookout man. Unfortunately, Fred was ostracized by the surviving officers of the *Titanic* for having revealed this omission (not very unusual for a ship on its maiden voyage when many obvious things are forgotten). In later years, he was down on his luck in the Depression, selling newspapers in his native Southampton. On January 10, 1965, two weeks after his wife died, he hanged himself from a clothes post in his garden. He was 86. Amid all the uncertainties and miseries of his life, one thing stands firm. If he'd had even a second-rate pair of binocs, he'd have spotted the berg much earlier, and the ship would have had plenty of time to swing clear.

However, an odd IF also applies. Suppose Fleet had not seen the iceberg at all, or had been slow in his reactions. Paradoxically, that would have been good!

IF (5) **the berg had been sighted a few seconds *later*, there would have been a head-on crash. The first compartment would have been smashed and flooded -- loss of life perhaps -- but with no danger of sinking.**

Trying to blame someone for the lack of binocs is hard. Suspicion might point to a number of weak links in the chain of supply from all the many, many people who outfitted the ship for her maiden voyage. It was a traveling town, almost a city. The White Star people and the navigating officers were disgusted with Fred for revealing the error. It was a pardonable omission at such a hectic time, but it was fatal.

The next IF hinges on the existence of the distress frequency of 500 khzs. (It was called kilocycles then -- it has since been changed to kilohertz, honoring Hertz, who came between Maxwell and Marconi in the development of radio). What means were available in 1912 to use this frequency and alert other ships that one was in distress? Some ships had two operators, some had one, some had none. As a result, there was no reliable way to contact many ocean-going ships in an emergency.

IF (6) **there had been a 24-hour watch by radio operators on all ships, serving in shifts to answer any distress calls coming in. . . .**

This IF loomed large in the mind of Marconi, who, at the time of the *Titanic* disaster, was waiting for her in New York. He had already seen the necessity for a radio auto-alarm to ring bells on all ships within range, to rouse a sleeping Sparks and get him to rush to the wireless room. Marconi had been experimenting, but the idea was not to become a reality until 1928. A series of four-second dashes on the Morse key (each separated by a single second of silence) would activate machines on all ships within 500-1,000 miles, to ring bells on the bridge, in the radio shack and in the R/O's cabin, to tell him to get on watch and listen for a distress message. Its universal use was made compulsory by international agreement 16 years later (1928). The auto-alarm is a wonderful device. It has saved thousands of lives. The potential importance of this auto-alarm idea for the *Titanic* concerns one ship, the *Californian*. She was at anchor 20 miles away all the time.

IF ⑦ the auto-alarm had been invented, and IF there had been one aboard the *Californian*, everyone on the *Titanic* could, and most likely would, have been saved.

But this IF hinges on scientific progress. Nobody can be blamed for not inventing something yet.

If we need to lay blame wherever it might legitimately rest, we must aim in the direction of the *Californian*. We'll have to concentrate on her Master, her Sparks, and her third mate.

IF ⑧ Captain Stanley Lord, skipper of the *Californian*, had headed for the *Titanic* immediately, perhaps everyone would have been saved.

"Immediately" means just that -- at the first SOS. However, this was impossible under the circumstances. Lord was -- still is to many -- the primary scapegoat. *Walter* Lord (no relation) has given him the old "one-two" by his books *A Night To Remember* (1955) and *The Night Lives On* (1985). In his defense, though, we'll see that he was utterly tired -- exhausted after 18 hours on the bridge, and therefore foolish, slow-minded. He couldn't have reached the *Titanic* in time, and most commentators haven't realized this fact. This will be carefully explained with diagrams later. Now let's deal with the other two crew members of the *Californian* mentioned.

IF 9 **the *Californian's* Sparks had been on watch 15 minutes longer, he'd have heard the *Titanic's* call for help.**

He wasn't supposed to be on watch; he'd already worked many hours of unpaid overtime. Still, that near-miss/maybe/ almost-but-not-quite aspect is infuriating.

There are three IFs left. They are concerned with the *Titanic*'s wireless department (two men), the *Titanic*'s chief mate and the *Californian*'s third mate. We'll get to them later.

For now, let us put the clock back to 9 AM, Sunday, April 14th, 1912, and then move it forward, hour by hour, to a few days after the *Titanic* sank. We'll live through each hour with the individual participants in the drama. Then, adding our hindsight wisdom and overall knowledge, we'll get the full picture.

***SUNDAY,* APRIL 14TH, 1912**

9 AM -- On board the *Titanic*. She's racing ahead at 22 knots. In the wireless cabin, Harold Bride, the junior operator, is on watch. He gets an ice warning from Cunard liner *Caronia*: BERGS, GROWLERS [overturned icebergs], FIELD ICE ALONG 42N from 49W to 51W.

Captain Smith acknowledges this. Bride's chief, Jack Phillips, is busy repairing a burned-out and grounded transformer secondary.

We have come to yet another IF, but this one is totally different. The others have all pointed to a possible prevention of the disaster. This could have made it worse. If Jack Phillips hadn't made this repair, he'd have been unable to send the SOSs that brought help. The 712 survivors probably owe their lives to his skill and persistence. Now that we've brought Jack on the scene, a thumbnail sketch would be appropriate, since he hasn't long to live. He is 25

John (Jack) Phillips
1887-1912
Chief Wireless Op. of the *Titanic*

(birthday four days ago); born in Godalming, Surrey, England. He became a "Marconi man" and is remembered by his Liverpool instructor as "a pleasant boy, well-spoken, good-tempered and friendly." He was last seen by his colleague Bride, walking across the tilting deck aft, minutes before the end. The memorial stone in Battery Park, New York, was originally suggested by a New York newspaper to perpetuate the memory of Jack Phillips only. A special memorial tablet to him is in a quiet, ivy-covered corner of the parish church in Godalming. Subscriptions from all over the world paid for it. It is well cared for, and its peaceful, green surroundings contrast with the grey, ice-chilled waters of the Atlantic in which he died.

10 AM -- Another warning, this one from the *Baltic*. (This ship had come to the rescue, three years earlier, when the radioman of the *Republic*, Jack Binns, sent out a CQD -- the distress signal soon to be superseded by SOS.)

10 AM to 6 PM -- Other ice warnings are acknowledged from various ships. The *Titanic* races ahead all day -- 22, 23, 24.5 knots.

6:30 PM to 9 PM -- The *Titanic*'s navigators plot a great rectangular field of ice ahead, 70 miles long, 12 miles wide, right across their course. No one is worried. They calculate, "We'll pick up to the ice by 11 PM."

Temperature has dropped from 50 to 32 degrees -- freezing. The sea is "like polished glass," all agree. There's no moon, but incredibly bright stars, "like diamonds."

10 PM -- Jack Phillips is in contact with Cape Race, plowing through a backlog of messages. As long as Phillips has been busy repairing the generator, Bride had been unable to deal with the dozens of messages to and from the passengers. He had been in brief contact with Cyril Evans, Sparks of the *Californian*, who was about 20 miles behind them.

IF 10

all the iceberg messages (six) intended for the *Titanic* had reached the chart-room (three never made it) perhaps the easygoing (careless?) attitude of the navigating officers would have changed. Since three did get through and were considered, I doubt that three others would have changed the officers' minds.

10:21 PM -- The *Californian* comes to a complete stop. The ice is a great wall ahead. Captain Lord goes into the radio room.

"Any ships around, Sparks?"

"Only the *Titanic*," says Evans.

"Well, tell them we're stopped by ice."

10:25 PM -- Evans starts calling the *Titanic*. He is doing this with a rotary spark gap transmitter (now extinct).

It had a broad note, sploshing around the 500 khz area, impossible to ignore. This would be very annoying to anyone like Phillips, busy with other things. That attention-commanding aspect of the sparks transmitter was in its favor for emergencies, and maybe it shouldn't have been completely outlawed.

Replying from the *Titanic*, Phillips is as curt as his normally good manners permit.

"PSE QRT OM QRL (Please shut up, old man. I'm busy)."

His rebuff isn't all that rude. Evans has been jamming his reception and Phillips thinks operator-operator chitchat is in the offing. Evans hasn't given any indication of an important message to follow. Cyril Evans (20 years old) on the *Californian* has been on watch since 7 AM, with a few breaks for tea and snacks. Otherwise he has been 16 hours on watch. His $20-per-month pay has been well earned. He has already worked plenty of unpaid overtime hours, and it will soon be midnight. He seems to be annoying Phillips by trying to speak to him. Wouldn't this be a damned good damned good time to go to bed? He decides to hover for a while, to try to catch Phillips again.

11 PM -- Up on the bridge of the *Californian*, Captain Lord and Third Officer Groves see the lights of a "medium-sized ship coming up from the eastward." Groves thinks it's a passenger ship, but Lord says, "No, it's a freighter."

> *It is the* Titanic, *of course. Lord is wrong. He's tired.*

11 PM to 11:35 PM -- Captain Lord watches while Groves keeps trying to communicate by signal lamp (blinker).

11:35 PM -- Evans, unable to get the attention of the *Titanic*, is fed up, tired. He decides to go off watch. He takes off his headphones, turns off the receiver, lies down on his nearby bunk, and falls asleep almost immediately.

> *There were no strict rules in those days about watch times. In five minutes' time, the iceberg will scratch the side of the* Titanic. *In exactly 25 minutes from now, an SOS will be coming out on 500 khz. The* Californian *receiver will be inoperative, and Evans will be asleep. Can he be blamed? Or is Phillips to be blamed for his rather gruff dismissal of Evans when he had tried to inform him about the ice? Even if that message had gotten through to Captain Smith, is it likely that he'd have slowed down, stopped, or done anything?*

Iceberg locations compiled from reports sent to U.S. Hydrographics office Sunday, April 14, 1912.

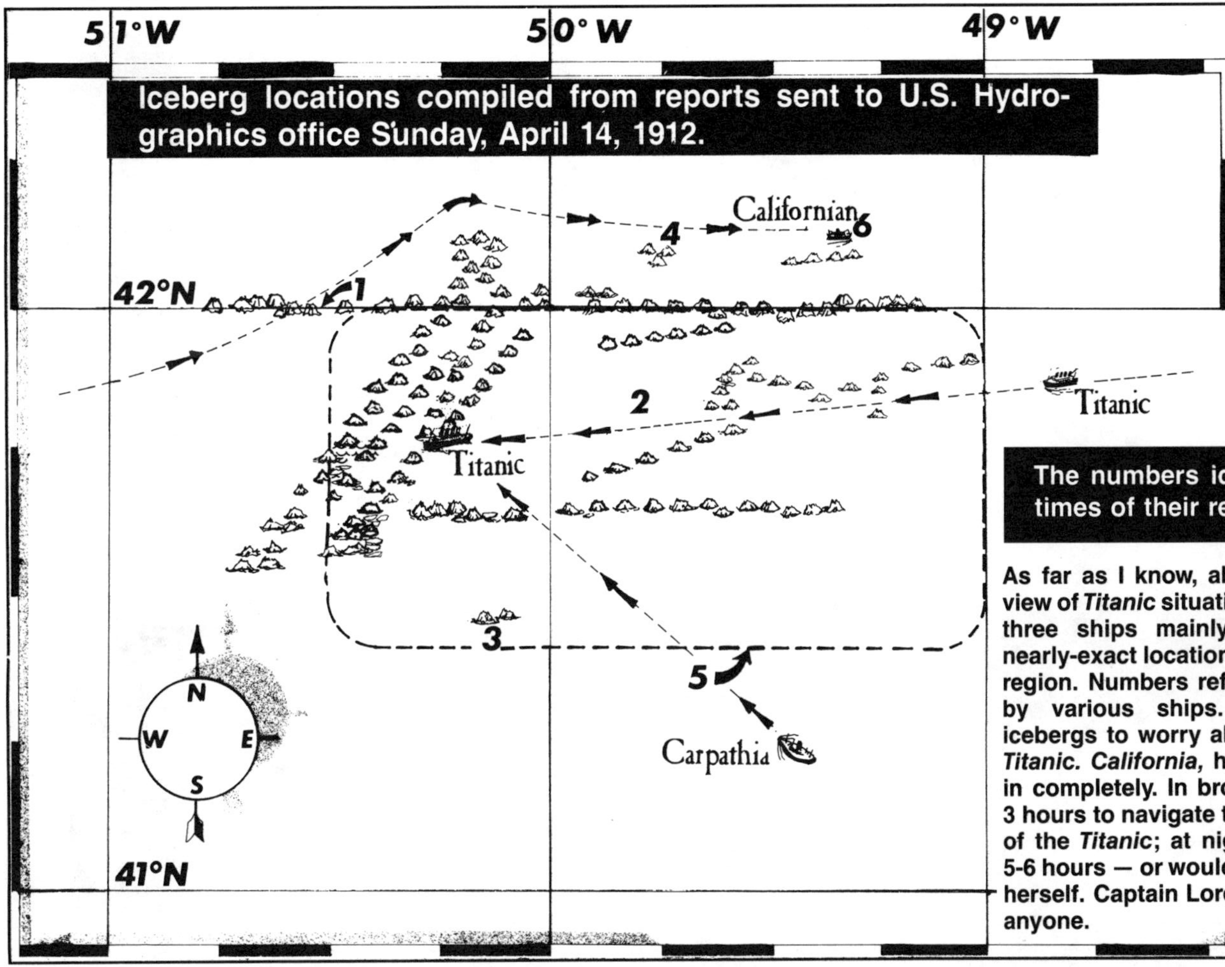

The numbers identify ships and times of their reports.

As far as I know, above is only birdseye view of *Titanic* situation which includes the three ships mainly involved, plus the nearly-exact location of the icebergs in the region. Numbers refer to ice reports sent by various ships. *Carpathia* had no icebergs to worry about until nearing the *Titanic. California,* however, was hemmed in completely. In broad daylight she took 3 hours to navigate the 19 miles to the site of the *Titanic*; at night, she'd have taken 5-6 hours — or would have been iceberged herself. Captain Lord couldn't have saved anyone.

THE ICEBERG REPORTS SENT TO TITANIC ON SUNDAY APRIL 14

Number	Time	Ship to Ship Message	Result
1	9:am	*Caronia to Titanic* Westbound steamers report bergs, growlers & field ice along 42N from 49W to 51W	Acknowleged by Capt. Smith
2	1:42pm	*Baltic to Titanic* Greek steamer reports passing icebergs & large quantities field ice 41.51.N 49.52W	Acknowleged by Capt. Smith
3	1:45pm	*Amerika* to US Hydrographic office via *Titanic* — Passed 2 large icebergs 41.27N 50.8W	Never got to *Titanic* bridge
4	6:30pm	*Californian to Antillian* (sistership) but intercepted by *Titanic* — 3 large icebergs to southward of us 42.3N 49.9W	Delivered to *Titanic* bridge
5	9:40pm	*Mesaba to Titanic* & all Ice report 42N to 41.25N, 49W to 50.30W — Much heavy pack ice & great number icebergs also field ice	Not delivered to Master or Mates
6	10:30pm	*Californian to Titanic* (tried to send to *Titanic*; was told by Phillips "Shut up/am working Cape Race"	No message sent or received

THREE OUT OF SIX GOT TO BRIDGE (SEE MAP NUMBERS, ETC)

ICEBERG MESSAGES

1. This one, at 9 AM, was acknowledged by Captain Smith.

2. He also acknowledged this one, then passed it along to Ismay, who showed it to some passengers. Ismay kept it until 7:15 PM, when Captain Smith asked for it so that it could be posted in the chart room. Look at the map on the previous page. Up to lunchtime, the *Titanic* had already been informed of the string of "bergs, growlers and field ice" along latitude 42N and the same stuff in position 41.51N 49.52W. The *Titanic* was heading right into it all at 24 knots or more, and darkness was coming on.

3. These icebergs reported by the *Amerika* at 1:45 PM were rather remote from where the *Titanic* was heading. This message never got to the bridge -- but so what? No matter.

4. This message was being sent to the *Antillian* by the *Californian*, but Bride, on the *Titanic*, intercepted it and took it to the bridge himself. When the *Californian* was about to repeat it for him, Bride said, "No thanks, I've got it." It was a very, very important message for iceberg location. Look at the map. The *Titanic* was heading right into an area that was quite clearly studded with icebergs along the latitude of 42N, with lots of them in the area just below latitude 42N and now one just above latitude 42N. Any further warnings about this area would be merely repetitious, which is exactly the way they were treated.

5. This message by the *Mesaba* reporting a large area as iceberg-covered never got to the bridge of the *Titanic*. The backlog of messages that hadn't been sent was growing. The time spent the previous evening in making repairs to the wireless had made the unsent message pile grow much more. But so what? The vital info had already been given to the bridge. Message 5 was nothing new.

6. Nor was this second message from the *Californian* ever received by the *Titanic* because Phillips irritatedly brushed Evans off. "Shut up. I'm busy with Cape Race." As Bride later explained, this would *not* be taken as rude -- Phillips didn't realize this was not a small-talk session in the offing -- he didn't want to be interrupted. But message 6 was nothing really new. Message 4 had been given to the bridge -- that was the important one.

Summing up, messages 1, 2 and 4 were quite adequate to show the situation. Number 3 wasn't too important. Number 5 was vague and repetitious, and 6 was nothing really new, considering the importance of 4.

The radio department's job was well done, looking at the map of iceberg positions reported.

11:40 PM -- *Titanic* lookout man, Fred Fleet, way up on the "crow's nest," sees a dark object looming up. He strikes three bells and turns a little handle which rings a bell in the wheelhouse. Sixth Officer Moody picks it up.

"Bridge."

"ICEBERG DEAD AHEAD!" yells Fleet.

"Thank you," from Moody, then relaying the information in a loud voice to First Officer Murdoch on the starboard wing of the bridge.

"ICEBERG DEAD AHEAD, SIR!"

Murdoch yells, "HARD A-STARBOARD!" as he bounds into the wheelhouse, and Quartermaster Robert Hickens, at the wheel, echoes the instruction and spins it clockwise.

In those days, starboard (right) referred to the stern (rear) of the ship. She was actually being put on a course to port (left). Until about 1924, all ships had tiller steering -- if you wanted to go to the left, you swung the tiller right. Now, as in a car, to go left you turn the wheel left.

By this time, Murdoch has reached the engine-room telegraph. He pulls it back and down as far as it will go, to FULL ASTERN. (It'll take some time before the ship can stop, let alone reverse, but this is the most violent braking action possible.) Then he yells "HARD A-PORT!"

Murdoch's reactions and actions were entirely natural and instinctive. Think about it. What would you do if you were threatened by imminent collision with an iceberg? Back off and turn to one side, right? Wrong! Reversing the engines and shifting the helm was the worst possible action! One or the other, but not both.

THE FIRST TEN SECONDS ...
ICEBERG DEAD AHEAD!
Up in crows' nest, AB Fleet yells alarm.
Cadet Gill quietly responds, then ...
Thank you Mr. Fleet.
... yells to ...
ICEBERG DEAD AHEAD, SIR!
Chief Mate Murdoch who shouts order to quartermaster
HARD A-STARBOARD!
runs into Wheelhouse and...
Switches engineroom telegraph to FULL ASTERN.
HARD A-PORT!

MURDOCH'S CHOICES

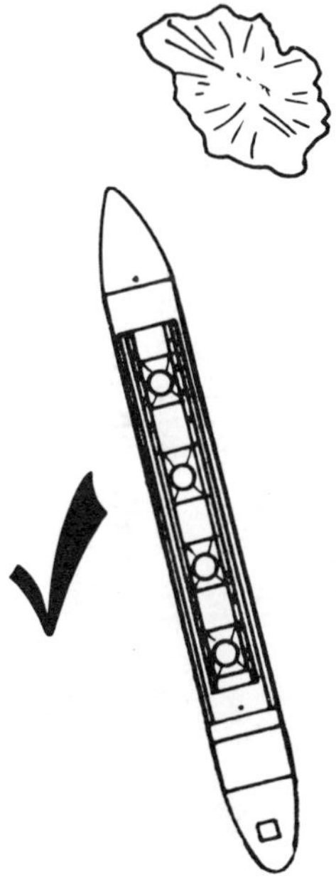

HARD A-STARBOARD!
Nowadays "Hard Right!" would be used. In 1912, this meant the ship would swing to the left (port).

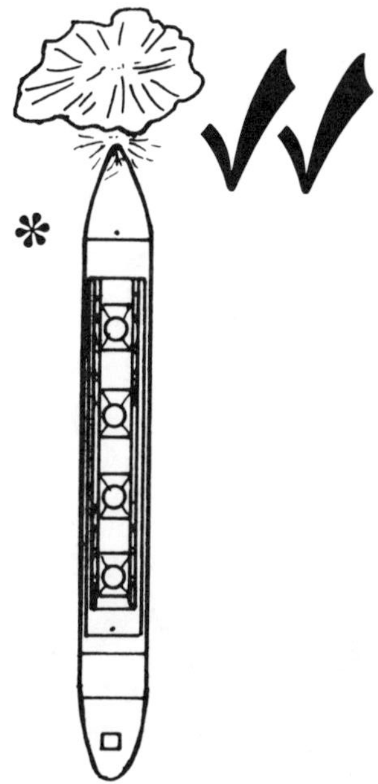

***Best possible choice!**
To keep a steady course, (straight into iceberg) signalling engine room "FULL ASTERN" ... OK. Bow smashed, but that's all.

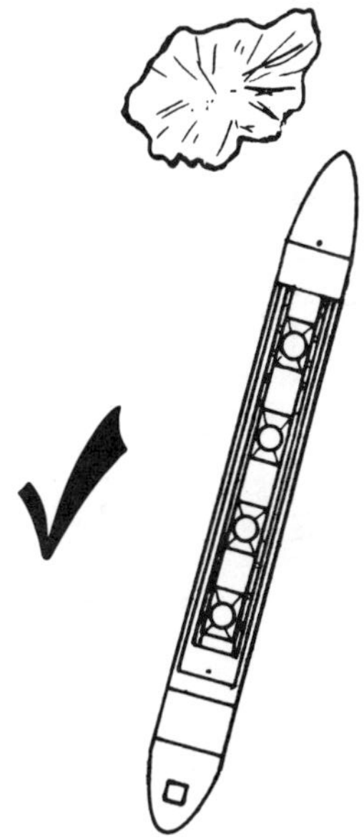

HARD A-PORT
Modern version "Hard Left!". Ship swung to *right*, (starboard). Sounds confusing! The orders applied to the rear of the ship. It had "tiller steering" as in a row-boat.

MURDOCH'S MISTAKE

In point of fact, Murdoch yelled "HARD-A-STARBOARD!" swung the telegraph to "FULL ASTERN" and then yelled "HARD A-PORT!"

Mixing three orders — no good. *Titanic* slowed up, and swung around slightly to be ripped open for 300 feet.

IF (11) **Murdoch had signalled FULL SPEED ASTERN immediately -- and *nothing else* -- the ship would have slowed down a bit by the time of the collision with the iceberg and then (as would have happened if the iceberg had been sighted later) the first compartment, the bows and all the front of the ship would have been badly damaged. Maybe some deaths and injuries, but that's all.**

Alternatively, to have veered off (to the right or left, no matter) at FULL SPEED might have pulled the ship clear. But she might just as well have run into the hidden (and larger) part of the iceberg. No, to have reversed the engines and accepted relatively minor damage was the best course of action. As it was, the Titanic slowed down, suicidally offering her whole port side to be ripped open.

Murdoch goes over to the electric switchboard and throws the switches to close all watertight doors. (They close, all right, but five compartments are about to be flooded.)

The time taken for all this (from the iceberg sighting to now) has been just 10 seconds!

Murdoch's wrong but quite natural reactions could only be offset by hindsight-wisdom-turned-into-foresight.

Did he commit suicide (shooting himself in full view of many passengers) as a boat was filling with people? It's very probable, because this rumor has different sources (people who were disconnected agree on basic facts).

He must have seen himself and his few seconds of undecided navigation as being the underlying cause of all the tragedy. Actually, his completely understandable error was only one of the IF clauses that would have averted the whole tragedy. (Any one of 12 would have succeeded.)

Officers and seamen stare at one another in tense silence as more seconds tick by.

11:41 PM (27 seconds later) -- A slight scraping impact is felt and a bow-scratching noise is heard.

The noise was identified later, but its source was not immediately ascertained on the bridge. The Titanic had (it was thought for 73 years) been ripped open by a 300-foot-long slash just below the water line. When she was inspected by Dr. Ballard and his crew in 1986, they found no slash or gash. Instead, the iceberg's impact had sprung all the bolts. However, the effect was the same. The Titanic *was doomed.*

11:45 PM -- The violent swing of the *Titanic*'s rudder has taken effect. She's now headed 5 degrees south of her original course, has slowed down, and will soon be still. Meanwhile, over on the *Californian* -- "Look, that ship's lights have gone out" (from Third Mate Groves).

"No, she swung left and stopped, I think. We can't see the lights from here" (from Second Mate Stone).

"Yes. She must have stopped for the ice, like us" (Captain Lord speaking).

There are quite a few people on the bridge. It's watch-change time. Second Officer Stone and Apprentice Gibson have come to relieve Third Mate Groves. Captain Lord is still here.

11:45 PM -- *Titanic* Junior Operator Bride awakes. He hasn't felt the mild shock of collision, but the throb of the engine has ceased, and the lack of noise and vibration awakens him. He decides to go right on watch, though he isn't due until 2 AM. Phillips seemed very tired, working all day on those repairs, so Bride goes into the radio room, and takes the phones from his partner. Phillips says, "We must have hit something. The ship's stopped." At that moment, the door opens, and in comes Captain Smith, bearded patriarch. He intends to retire after this voyage.

"We've struck an iceberg, and I'm having an inspection made to tell what has been done to us. You'd better get ready to send out a call for assistance, but don't send it till I tell you." He hurries back to the bridge.

Bride and Phillips look at each other, then start laughing.

None of this account is fanciful. It is based on later accounts, sometimes verbatim. Bride survived. This is his account to the New York Times.

They are on the world's biggest, strongest, safest ship. It's called "unsinkable." The weather is very calm. So they've been scratched by an iceberg? Watertight compartments have been designed to take care of that. The Old Man is just a fusspot, but that's part of his job.

12 Midnight -- Captain Smith is back at the door of the radio room. "Send out the call for assistance," he orders. "Here's our position." He puts a slip of paper on the desk.

"Which call, Captain?" asks Phillips.

"The regular international call for help." The Old Man hurries away. Phillips is no longer laughing. He switches to full power, hits the key, "CQD CQD CQD DE MGY MGY MGY."

"Better use that new SOS signal," says Bride.

It had been adopted by international agreement in 1908, but CQD still lingers for Marconi operators.

Phillips starts again: "CQD CQD CQD/SOS SOS SOS DE MGY MGY MGY HAVE STRUCK ICEBERG WE ARE BADLY DAMAGED TITANIC 41.46 N 50.14 W MASTER."

MGY was the coded designation for the Titanic. *Three-letter combinations were used for ships. Nowadays, they are assigned to coast stations only. Ships have four letters, or letters and figures.*

12 Midnight still -- On the *Californian*, Third Mate Groves has been relieved by Stone and is going down the stairway to his cabin and bed, thinking about that mysterious ship he and the Old Man have been watching for half an hour. On an impulse, he turns to the radio shack and enters.

"Excuse me, Sparks."

Evans is very near sleep. He replies with a querying grunt-- "Uh?"

"What ships have you got, Sparks?"

"Only the *Titanic*," mutters Evans, and buries his head in the pillow again.

We now come to the most tragic IF of all.

Groves, backing away, goes into the radio shack. He can read Morse "if sent slowly." He picks up the headphones from the desk by the receiver, and puts one earphone on his head. He hears nothing.

Of course not. This ancient type of receiver was the old magnetic type, needing to be wound like a clock and carefully adjusted.

At this very moment, CQD and SOS coming out, full strength, from the *Titanic*, are heard by Cape Race, 379 miles away. The *Californian*, only 19 miles away, has Groves with a dead earphone to his ear. Evans is asleep.

IF (12) **the *Californian* receiver had been working now, 1,517 people *might* not die. (See summary of all IFs near the end of this chapter.)**

12:05 AM -- Phillips has just had his CQD/SOS acknowledged by the German SS *Frankfurt*, 153 miles to the southwest. We switch now to another ship, the SS *Carpathia*, who did *not* hear the distress call. Her Sparks, Cottam, shouldn't be on watch at all, but he wants information on a British coal strike, so he has tuned in Cape Cod. The Cape Cod traffic list has "MGY" on it. Cottam decides to call the *Titanic* and tell her Sparks there was a message for her at Cape Cod.

No trouble to raise MGY, but Cottam is startled by what Phillips says: "QSL? MY CQD/SOS OM? (DIDN'T YOU GET MY CQD/SOS, OLD MAN?) COME QUICKLY WE HAVE STRUCK A BERG!"

The Carpathia *is to be the only rescue ship. Her Sparks, Cottam, would normally have been asleep. For an irrelevant reason -- to get news about a strike -- he happened to be on watch.*

He rushes up to the bridge, to First Officer Dean, with the news. They both run down to the captain's cabin, burst in without even knocking. Rostron, a stickler for discipline, is about to reprimand them, but Dean blurts out, "Cap'n -- the *Titanic* is in trouble -- hit an iceberg -- she wants help!"

Rostron rolls out of his bunk. "I'll be up in the chart room in moments, Mr. Dean." The chief officer disappears.

Rostron fumbles for his pants, shoes, socks. "Mr. Cottam, are you absolutely sure it's the *Titanic* and she requires immediate assistance?"

"Yes, sir."

"Are you absolutely certain?" (Dammit, "unsinkable," calm, clear night--all the usual objections.)

"Quite certain."

"All right, tell 'em we're coming as fast as we can." Up in the chart room, he sets a new course. Northwest 52. There's 58 miles to go. At their usual 14 knots, they'll take four hours ... too long. He phones the chief engineer.

"Pour it on, Chief. Call out the off-duty watch. Cut off heat and water. Pour every last spoonful of steam into the boilers."

He tells First Officer Dean, "Rig electric clusters along the ship's sides. Open all gangway doors. Rig chair slings for the sick and injured."

Boats are swung out, ladders rigged, blankets, bos'ns chairs -- all are readied.

He tells the ship's surgeon, Dr. McGhee, to set up first aid in each dining saloon. He orders the chief steward to have plenty of soup, coffee, tea, brandy ready for the survivors.

This Old Man is leaving nothing to chance. Extra lookouts, emergency gear (including bags for hoisting kids aboard), even heavy chairs, with ropes attached, to bind securely any survivors who have gone mad!

"Systematic completion" was to be the praise-laden summary by Senator Smith at the U.S. Inquiry later. Captain Arthur Rostron, 27 years at sea, seven years with Cunard, two years a skipper, and only two months on this old tub, is having his first real test. He's justifying his nickname of "Electric Spark."

The passengers -- 150 first-class (elderly Americans on cruise to the Med and back) plus 575 steerage (Italians and Slavs going home)--are soon aware of an emergency as the vessel shudders under the strain. Down below, anyone who can hold a shovel is piling on the coal. The rickety old ship knifes ahead 14 .. 15 .. 16 .. 17 knots.

12:15 AM -- *Californian*. Second Mate Stone, who has just relieved Groves, is listening to Captain Lord.

"Watch that ship over there, Stone. I'm going to lie down in the chart room." Lord, with 18 hours on the bridge and no sleep for two days, is very, very tired.

This is the occupational hazard of every captain. He has dictatorial powers at sea. He pays for it by being responsible for everything. Even if he's nearly asleep on his feet, he cannot delegate authority. Before condemning him for the way he behaves, we must reconsider those facts -- 18 hours on duty, 2 sleepless days -- and be surprised only if there had NOT been faulty judgment and a wrong scale of importance.

12:20 AM -- *Titanic*. Crewmen start lowering lifeboats. Down below, the engineers have reduced boiler pressures, to prevent an explosion, and this makes for a shrill noise of escaping steam. This is harassing to Phillips, struggling to detect faint radio signals.

Crude crystal detectors were all they had in 1912. At their best, they were weak.

12:33 AM -- *Titanic*. Phillips hears from the Carpathia: "WE ARE BUSTING OUR BOILERS TO GET TO YOU ETA 0400." Down on the promenade deck, eight musicians start playing ragtime.

12:45 AM -- *Californian*. "I'm going to lie down in my cabin, Mr. Stone," says Captain Lord. "We can't move by this ice until morning. But watch that ship, do you hear? Call me if anything seems odd."

12:45 AM -- *Titanic*. The first rocket is sent skyward. (It is seen by the *Californian*, but not associated with distress. They think they're company signals.) Boat number 7 is almost ready for launching. Ismay calls, "Gentlemen, please stand back. Put in brides and grooms." In go 25 passengers and two crew members. First Officer Murdoch calls, "Lower away!" Down they go--down, down, 70 feet.

Number 7 lifeboat enters the water with only 27 people, though its capacity is 65. Number 1 boat has only 12 out of a possible 40. Women and children are almost the only occupants.

Colonel Astor places his young bride in a boat and rejoins the men. This and other acts of heroism are tinged with a slightly mocking air of unreality. The *Titanic* isn't really sinking (says the general mood). All this fuss is just precautionary.

Boat number 5 is next. Fifth Mate Lowe is annoyed by Ismay, who is shrilly yelling, "Lower away! Lower away!" Tells him to "get the hell out of the way."

"Any more before this boat goes?" calls Ismay.

"I'm only a stewardess," says one girl shyly.

"Never mind. You're a woman. Take your place." Chief Murdoch orders Pitman (sixth mate) to get in and take charge. The boat lowers to the water. It could hold 65, but only 35 are aboard. The loading officers are worried that, if loaded to its full capacity and with so far to fall, it might buckle if fully loaded.

12:55 AM -- Boat number 6. Captain Smith finds out that Major Peuchen is a yachtsman and orders him in to take charge. Then he bawls down to the lifeboats from the wing of the bridge: "Row toward that light!" (It is the *Californian*.)

The efforts of the engineers 122 feet below the boat deck, pumping out thousands of tons of water, undoubtedly prolonged the life of the ship, at the expense of their lives. They all died down below. But Jason Jr. (the roving camera used in 1985) did not see their bones. Mollusks (marine scavengers) had eaten them up. They worked on the pumps and maintained the electric power for lights and radio until the very end.

1:00 AM -- Boat number 3. Four first-class men and some firemen get in. Clumsily, with little coordination, the boat is lowered to the water.

1:05 AM -- Boat number 8. Twenty-five women are in it. "Any more ladies?" bawls Captain Smith.

Mrs. Strauss refuses to get in, as an officer hustles her. "I'll not leave my husband," she says. "We've been together all these years, and I'll not leave him now." She stands by, waves her handkerchief to the others as the boat goes down.

"Strength and calm showed in their faces," said one of the passengers later. Mrs. Strauss is one of the true legends. Her being Jewish in an era of anti-Semitism gave a poignancy to her sacrifice. William Jennings Bryan compared her to Ruth in the Old Testament ("Entreat me not to leave thee"). The president of the Studebaker Corporation suggested that her name be the first on a proposed engraved list of all the dead on the Titanic.

1:12 AM -- Boat number 8 goes down. It has seven crewmen, three first-class men, and one first-class lady with her maid--12 altogether.

1:15 AM -- *Californian*. Stone blows on the speaking tube to the captain's cabin. "Cap'n, that steamer is moving off to the south, and she sent up a rocket."

"Was it a company signal?" asks Lord.

In 1912, they used colored balls and colored flames as signals between passing ships.

"I don't know, but they look white."

A white rocket -- any rocket -- is used for distress.

"Use the signal lamp," orders Lord. Stone relays this to his apprentice, Gibson.

"Get out on the bridge, and keep flashing the Aldis lamp until they answer you."

1:20 AM -- *Californian*. There has been no response from the *Titanic*. Stone blows on a speaking tube to the captain's cabin but gets no reply. He looks at Gibson, shrugs his shoulders. "We're both getting nowhere. The Old Man is in a deep sleep. Look, there's another rocket!"

Gibson points. "Look at her now. Don't the lights seem queer?"

1:20 AM -- *Titanic*. Number 9 boat is down.

1:22 AM -- *Titanic*. Number 10 boat goes down. The *Titanic* is listing. There's a three-foot gap from boat to rails. A French lady jumps--nearly misses, holds on by her hands until pulled in.

1:25 AM -- In the radio shack, Phillips talks to sister ship *Olympic*: ARE YOU STEERING SOUTH TO MEET US? WE ARE PUTTING THE WOMEN OFF IN BOATS. Lots of boats now going down. Number 11 (starboard side) is lowered to the level of A deck, where 24 third-class women are waiting. It ends up with 70 people--the most in any boat.

At number 13, a last call for women goes unanswered. Five men jump in.

Number 14. Lowe comes over to the port side and takes charge. He spots a boy dressed as a woman, threatens him with a revolver and makes him get back on the *Titanic*. He fires the gun along the side of the ship to discourage more people entering an already overcrowded boat.

"Women and children first. Men stand back," calls Lowe. There's no distinction as to class or nationality.

◐ ● ◑ 1:30 AM -- SS *Frankfurt*: ARE THERE SHIPS AROUND YOU ALREADY? Phillips ignores this, but the *Frankfurt* is persistent: REQUEST MORE DETAILS. Phillips (exasperated): STAND BY KEEP QUIET.

◐ ● ◑ 1:30 AM -- *Titanic*. Thomas Andrews, the man who designed and built the ship, is doing all he can to prolong the life of his ship and save what human lives he can. Down in the engine room, he toils away with Bell (the chief engineer) and 35 licensed engineers, just to get power for the lights and the wireless.

Andrews and all the engineers perished.

All the lights of the *Titanic* will be blazing and the wireless gear fully operative until within a minute or two of the final plunge. The engineers will keep the power coming up.

There is a monument to them all in Southampton. Andrews came up on deck at the end, and was last seen lashing deck chairs together and throwing them overboard. They'd be some use to struggling victims in the water. Most of these accounts were written at the time or shortly afterward -- from ships' logs, diaries, Court of Law hearings, etc. They are accurate, reliable. Some anecdotes, though authenticated by various witnesses, cannot be pinned to an exact time during the last two hours and 40 minutes of the Titanic. *There are so many, there has to be selection and elimination. Who can resist the story about the Titanic waifs? Two children, about two and three years old, survived in one of the boats. Their names were Lolo and Momon Navratil, according to Walter Lord, in his sensational and historically accurate* The Night Lives On *(1986), sequel to* Night to Remember *(1955). He says, "Their father had kidnapped them from their mother and was taking them to America under an assumed name to start a new life. He put them in the last lifeboat, stepped back and went down with the ship. The children were too young to know who they were, and their identity remained a mystery for days." "For many years" would be more accurate. Where and how Walter Lord got his info doesn't matter. The mystery has at last been cleared up.*

Eight musicians played ragtime as the ship went down, to maintain the courage of the doomed passengers. This went right to the hearts of the whole world. All the members of the band perished.... The shabby buck-passing tale of the way the musicians' families were treated is revealed by Walter Lord.* Who'd be responsible for them? Sorry, said the White Star Line, they were second class passengers. Sorry, said the employment agency, you must go to the insurance company. Sorry, said the last-named, they were independent contractors. Sorry, said the judge who was appealed to --they weren't employed by anyone....

Well, eventually the families were rescued by the Titanic Relief Fund.... When bandleader Wallace Hartley's body was retrieved from icy Newfoundland waters and sent to his home town Colne in Lancashire, England, a half-mile procession followed his casket. Boy Scout buglers played the Last Post; instruments of the original scene were lilting violins on the sloping decks of the *Titanic.*

1:35 AM -- Boats number 15 and 16 go down from the stern of the boat deck, 60 people in each.

Detonators have been going off for nearly an hour, firing rockets. Now they're finished. The starboard side is clear and there are no passengers in sight.

"Any more women and children?" calls Ismay. No answer. He and a first-class male passenger leap into a collapsible (one of the makeshift lifeboats).

1:46 AM -- *Californian*. "That's the eighth rocket," said Third Mate Stone to his apprentice.

"Shall I go down to tell the Old Man?" (from Gibson). "You don't seem to be able to rouse him on the tube."

"Maybe. Wait awhile."

1:47 AM -- *Titanic* (Radio shack). Phillips to Cottam (*Carpathia*): *COME AS QUICKLY AS POSSIBLE, OLD*

**The Night Lives On* (Wm. Morrow & Co., 1986)

MAN. ENGINE ROOM FILLING TO BOILERS.

Bride drops his overcoat over Phillips's shoulders and goes out on the deck to see the situation. The forward well-deck is awash -- water rippling around cranes and mast, splashing against white superstructure. The deck is steeply slanted. There's an ugly list to port. Chief Officer Wilde: "Everyone to starboard side to straighten her up!" All obey. *Titanic* is on even keel again.

Lightoller finds number 2 boat full of men. Yells "Get out, you damned cowards!" Boxhall (fourth officer) is in charge, appointed by Captain Smith. He has three crewmen, one steerage man, and 25 women -- 29 in all.

1:49 AM -- Lowe is in charge of boat number 14 as she sinks past A deck. "Who's next for the boat?" he calls. No answer.

"They didn't seem to want to go," he told the American inquiry.

Captain Smith is talking on the speaking tube from the bridge to the radio room.

"The ship won't last half an hour, Mr. Phillips."

Phillips and Bride go to their adjoining room, dress warmly, don their life-belts. They are sitting thus when Captain Smith appears at the radio room door for the last time.

"Men, you have done your full duty. You can do no more. Abandon your post. Now it's every man for himself."

1:50 AM -- Number 4 has 14 women. The *Titanic* is very much down by the head at an increasing angle. Three crew members swim to boat number 4 and climb in, saying the ship will be under in minutes.

1:55 AM -- Phillips is still working. The *Olympic* has just signalled, "LIGHTING UP ALL POSSIBLE BOILERS AS FAST AS WE CAN," but the *Carpathia* is much nearer.

Phillips asks Bride to go out on deck, see if all the lifeboats have gone. Bride does so, and sees water coming over the boat deck, which is normally 60 feet above the water line. There are no lifeboats visible.

◐ ● ● 2:00 AM -- *Titanic*. Bride reenters the radio room, tells Phillips the news. Phillips, back in contact with the Carpathia, sends his last message: COME QUICK ENGINE ROOMS FLOODED TO BOILERS.

Thanks to the engineers, he had full power still available. It hadn't been necessary to switch to the battery-powered emergency transmitter.

◐ ● ● ◷ 2:03 AM -- *Carpathia*. Cottam is spied by a stewardess, hunched over the set--only wearing his shirt. It was cold. He'd started to undress at the first CQD/SOS, and he hadn't put his coat on again. The stewardess finds it, puts it over his shoulders. Cottam thanks her, but doesn't move.

◐ ● ● ◷ 2:05 AM -- *Californian*. Second Mate Stone says, "A ship doesn't fire rockets at sea for nothing. Go and wake the Old Man. Tell him we have seen eight rockets."

Gibson does as he's told.

Captain Lord drowsily mutters, "Were they all white?"

"Yes, Captain."

"What time is it?"

"Two-oh-five AM, sir."

There's no further reply. Gibson waits, but Captain Lord has dropped off to sleep again.

A Moment of Truth -- with Sheer Fatigue as the master of the Master! As well lay blame on Gibson -- a cadet, a nobody, the lowest of the low -- for not daring to lay hands on the Master of the ship and yell, "Wake up, sir! Wake up! We think a ship's in trouble!" Instead, Gibson rejoins Stone on the bridge.

◐ ● ● ◷ 2:08 AM -- A group from steerage stop at gate labeled "Women and children only." Mr. Goldsmith embraces his wife and son, who then go through. Mrs. Goldsmith is given a ring by a male friend. "If I don't see you in New York, give this to my wife." Alfred Rush, who had had his 16th birthday a few days before, jerks his arm away from the steward who had pulled Mrs. Goldsmith and her son through. "No," he says, "I'll stay here with the men."

2:10 AM -- Lightoller had all the men's arms linked in a semicircle around collapsible D, to let the women and children through. Chief Officer Wilde comes over from the starboard side.

"You go with her, Lightoller."

"Not bloody likely," he replies and jumps past to the davits. Two men see the collapsible come alongside and no one in the bows. They tumble in.

2:12 AM -- Phillips, still hunched over the set, is too absorbed to see a stoker who had come into the radio shack. Bride sees him gently unfastening Phillips' life jacket! Bride leaps at the stoker. Phillips jumps up. All three begin to wrestle. Bride pins the stoker's arms to his sides, and Phillips hits him again and again. Eventually he's knocked out and limp in Bride's arms. Then they hear the sea gurgling up to A deck.

"Let's get out!" yells Phillips. Bride lets the stoker fall, and the two radiomen run out onto the boat deck.

Let Bride continue the story, which is part of the electrifying five columns on the front page of the New York Times *(for which they paid him $1,000): "I regret to say we left too hurriedly in the end to take the man with us. Without a doubt he sank with the ship in the Marconi cabin as we left him. I now assisted in pushing off collapsible D, which was on to the boat deck. Just as the boat fell, I noticed Captain Smith dive from the bridge into the sea."*

Bride watches Phillips go aft. He never sees him again.

The fate of Phillips is vague. He was seen clinging to one of the collapsibles. Some say he was seen on the collapsible. He probably died en route to the Carpathia;

but there's no evidence as to how/when he died, nor what was done with his body.

2:15 AM -- Bandmaster Hartley continued playing; "Autumn" flows out across deck and water.

The first legend about the tune played being "Nearer My God to Thee" was started by a Canadian passenger, Mrs. Dick, in her story to a reporter. It seemed apt. This was discounted by Harold Bride, who said it was "Autumn." It was thought he was talking about a hymn called "Autumn," which is robust in mood and melody -- just the reverse of the other, which is the essence of submission to God and fate. But he was most probably referring to a music-hall tune called "Autumn," popular at that time in England.

The *Titanic* plunges forward. An immense wave comes over the bridge and boat deck. Collapsible B overturns, but Harold Bride keeps his head as the *Titanic*'s bow dips and the stern comes up, up--the great wave rolling across the deck. He grabs an oarlock of collapsible B. He and half a dozen others are all washed off together, beneath the upside-down boat.

Charles Lightoller dives off toward the sinking foremast. Far astern, Father Thomas calls, "Prepare to meet God!" and a crowd of emigrants fall to their knees, chanting the Rosary.

On A Deck, there were, until a short while ago, gentlemen who, with complete aplomb, were still playing cards on slanting tables. Liquor had been available for all. Paddy Dillon, from Belfast, downs three slugs as the ship lurches forward. He stuffs a brandy bottle into his hip pocket and falls overboard. (He's going to survive, helped in no small way by that brandy.)

On the boat deck, Colonel Archibald Gracie sees a mass of humanity surging onto the deck, up from steerage. There are a lot of women. He thinks, I thought they'd all gone in the lifeboats. As the huge wave comes closer, Gracie leaps up with it as if surfboarding, and it carries him up to a sort of safety. Away from the ship, he is whirled around, swims under the water, fearful that "the hot air coming up from the boilers might boil me up." He later finds out that the second officer had had the same feelings. He swims a good distance from the ship -- how far he doesn't know -- but finds wreckage and bodies all around.

From boat number 14, the ship looks like an enormous glowworm alight from end to end, electric lights blazing from every cabin. Lights on deck, lights on the mastheads. Everyone hears the band playing "Autumn."

Huddled survivors in the boats watch the last moments of the *Titanic*. She sinks lower by the head until almost perpendicular--lights still bright. (Those fabulously heroic engineers!)

Collapsible B is overturned. Bride is underneath. Then the forward stack collapses. The lights all go out. They come on again for one second, then go off forever.

An account later says, "We could see almost 1,500 people still aboard. In bunches, only to fall apart in masses or singly as the great after-part of the ship -- 250 feet of it -- rose into the sky.... "Her black hull against the stars was like an enormous finger."

She is now perpendicular and motionless. The spectacle of her three massive propellers towering against the sky is awesome. She hovers in this attitude for a full five minutes, while the massive engines, 29 huge boilers, fittings -- *everything* on the ship from grand pianos to anchor chains -- careen down into the bow, making the most awful noise the survivors will ever hear. The *Titanic* corkscrews to port, then settles at a 70-degree angle.

Since the lights were on until the Titanic *upended and the dynamos tore loose to crash in the bows, the inner hull held together without shearing her riveted longitudinal connections. Otherwise, the wires would have stretched and snapped. The lights would have gone out earlier.*

In one boat Frank Goldsmith (nine years old) sees a "column of black smoke rising up and flattening at the top like a mushroom."

As a symbol of utter dreadfulness, it is comparable to the appearance of the first atom bomb 33 years later.

All is quiet now. Nothing more is falling. The ship slips under. Quietly. In collapsible A, August Wennerstrom is very close by, and confirms what others, including Harold Bride, say about there being no suction. She goes under so slowly.

As her flagstaff disappears, there's a small ripple and a sound "like a small boat running off a beach." Thus the biggest ship in the world commences her two-mile dive to the seabed. The time is 2:20 AM, April 15th, 1912.

2:20 AM to 3:00 AM -- Now, at last, the ether (the medium for electronic waves) is silent. The ships who responded to the SOS -- the *Mount Temple*, the *Frankfurt*, sister ship *Olympic* and, above all, the *Carpathia* -- are racing toward 41.46N 50.14W in a silence that contrasts strongly with the questions and answers, pleas, grim facts and just chatter, all of which have been coming from the *Titanic* for two and a half hours.

But within a quarter-mile radius of the spot where the *Titanic* sank comes a dreadful sound. It is heard by the hundreds of survivors in the lifeboats all around.

The noise is the sound of a thousand dying people. Most have life-belts bearing them up. Icy water is killing them. For 40 minutes, their pleas for help will assail the ears of those in the boats. For 40 years and more after all this, they'll echo in the minds of the survivors. Those who are far off cannot help. Those who are near dare not do so, for fear of overcrowding and sinking their boats.

* * * * *

We'll now stop using the present tense. The Titanic has gone. We must evaluate, judge, philosophize, reach back in time.

At the American inquiry, nearly a week later, Senator William Alden Smith was relentlessly questioning Sixth Officer Pitman. It had been established that, immediately after the foundering of the Titanic, *there had been a "long, continuous wailing chant" coming from 1,000 people in a 200-foot circle.*

"There was a continuous moan for about an hour," said Pitman.

"And you lay in the vicinity of that scene for about an hour?"

Pitman, who had already been showing signs of distress under Senator Smith's probing, broke down completely:

"Oh, please, sir. I can't bear to recall it. I wish we might not discuss the scene."

Smith said softly, "I have no desire to lacerate your feelings. But we must know if you drifted in the vicinity of that scene for an hour.

"Yes," sobbed Pitman.

"Did these cries of distress die away?"

"Yes. They died away gradually."

* * * * *

2:20 AM to 3:15 AM -- What had happened in boat number 5 (which Pitman commanded) was: hearing the cries, he said, "Now, men, we'll pull toward the wreck." But the women had objected.

"Why should we lose all our lives in a useless attempt to save others?" Pitman was argued with, and shouted down. The noes were not only numerically superior; they lay on the oars, to prevent them being used. There were 40 people in a boat which could hold 65.

(In boat number 2) "Shall we go back?" asked Fourth Officer Boxhall of the ladies. They said "No!" And the boat -- 60% full -- also drifted while that moan went on for nearly an hour. In boat after boat, it was the same story.

Two boats differed from the rest. Number 4 (Quartermaster Walter and Mate McCarthy in charge) had picked up three men from the water before the Titanic went down. Later, they started rowing back to those floating and swimming. Several women objected so strongly that they lay on the oars to prevent the men rowing. In spite of that, the rowers kept going. When those in the water saw the boat coming, some started swimming toward it. Five men were pulled aboard, including Paddy Dillon with his bottle of brandy. Two of the other four died before they reached the *Carpathia*. In the meantime, they had pulled alongside number 14 (Lowe's boat).

Lowe had under his command boats 14, 12, 10, collapsible D, and now 4. He distributed the 58 occupants of number 14 into the other boats -- with some protests -- and ordered the mast erected, though there was no wind for a sail, yet. He decided it wasn't safe to take the boat into the mob of yelling, moaning people--swimming or just floating. He cold-bloodedly waited "for the people to thin out" (bluntly, to die off).

One watercraft (and it hardly earns the dignity of even that title) was collapsible B. With no means of self-propulsion, it wasn't a real boat, and anyway it was floating upside down. Those on it certainly couldn't have returned to pick up others, but it remained a focal point for dozens struggling in the water and had quite an interesting crowd. Harold Bride had been there -- underneath -- from the beginning. Second Officer Lightoller scrambled aboard. Then came a Yorkshire justice of the peace, with a shaggy fur coat. Next, Colonel Gracie, one of the last on the *Titanic*. He had been on a plank, then a wooden crate, and now this. (The collapsible was sinking lower into the water with each newcomer.) Assistant Cook John Collins got on next. Then Bride managed to get away from underneath, and got on top, helped by the others (he could hardly move his frozen legs). The rickety boat/raft now had 30 aboard it. Swimmers were discouraged from further boardings. "One more will sink us all!" cried out someone. One swimmer was cheering the others on with a voice of authority: "Good boys ... good lad!" But he didn't ask to climb aboard. Walter Hurst held out an oar to him, but the swimmer was too weak to seize it. He spun around in the water and sank. Walter Hurst always maintained it had been Captain Smith himself.

Collapsible B was sinking a little lower every minute. Lightoller took charge on one end, which might be euphemistically called the "bow." Seeing Harold Bride at the other end (or "stern"), he shouted to him, asking what ships were on the way. Bride called back, "*Baltic, Olympic* and *Carpathia* -- *Carpathia* nearest -- probably see us about daybreak." This cheered up the others a bit.

One occupant of B was very lucky. Chief Baker Charles Joughin had been awakened by that grinding jar as the ship hit the iceberg. He had been pretty busy for the next two and a half hours. When he wasn't carrying loaves (as food for those in lifeboats), he had been literally throwing hesitant women into the boats, or throwing deck chairs overboard as rafts for anyone to hang onto. Every now and again, he'd pop back into his own cabin for liquid comfort--nips from a bottle of whiskey. When the *Titanic* was about to sink, he had been on the stern. As the ship went under, Joughin had just stepped off into the water without even getting his head wet. Warmed and insulated by all the whisky he'd had, he swam around for an hour and a half. Then, by the gray light of dawn, he saw and was picked up by collapsible B.

3:00 AM to 3:50 AM -- Where the *Titanic* had sunk was an area of complete silence. There was no moon, only starlight. None of the dead bodies dangling from life jackets would be seen by anyone until daybreak. The lifeboats were rowing toward the lights of the *Californian*, where the disappearance of the *Titanic* lights was the sole topic of discussion.

The *Frankfurt, Olympic* and the *Carpathia* were all heading for the spot where the *Titanic* used to be. On the bridge of the *Carpathia*, particularly, all eyes were straining ahead. They expected to see survivors any moment.

Over a radius of four or five miles, the 18 boats wandered or drifted. In boat number 14, Fifth Officer Lowe had been waiting an hour and a half for the swimmers "to thin out." He decided to look through the mass of floating bodies for any survivors.

"There were a good few dead," said Seaman Edward Buley, oarsman of boat 14, "but they hadn't drowned. They were frozen." They found only three people alive. Then Lowe rejoined the boats he'd left drifting -- numbers 12, 10, 4 and collapsible D. A breeze came up, and they hoisted a sail. The *Carpathia* was visible to them now, and they started to row toward her.

Collapsible B had been very hard to keep afloat, with all standing on her bottom. Two had died and fallen off (Jack Phillips may have been one of them). Harold Bride was still standing, though his feet had been frozen in the water. "Finally," said Colonel Gracie, "dawn appeared, and there on the port side, glory be to God, we saw the steamer *Carpathia* about four or five miles away. The other *Titanic* lifeboats were rowing toward her."

4:00 AM -- The *Carpathia* stopped. A lifeboat had been spied. It was boat number 2, Fourth Officer Boxhall in charge. He shouted up to the bridge. "We have only one seaman and can't work very well."

"All right," shouted Captain Rostron and nudged the *Carpathia* closer until Boxhall called out, "Stop your engines!"

One boatload after another, they arrived and were taken aboard the *Carpathia*. Many boats were only partially filled. "Unsinkable" was the word linked to the *Titanic* from her launching. Lloyds gave her a special, low insurance rate. "God himself couldn't sink this ship," a steward told a passenger. But why pile up the ironies? Four hundred sixty-seven spaces in the lifeboats were unfilled. That many lives were needlessly wasted.

When number 7 boat arrived, Mrs. Dodge's little boy was hauled up in a mail sack. A steward rushed up with coffee, but Master Dodge announced, "I'd sooner have cocoa!" He got it.

Collapsible D had Bruce Ismay. Dr. McGhee offered him soup, or something to drink. "No. I don't really want anything at all," said Ismay.

"Do have *something*."

"If you can get me to a room where I can be quiet, I wish you would." So that's where he went; and stayed, for the rest of the trip to New York. Nothing to eat. No visitors. Only opiates to help him sleep.

By 8:15 AM, all boats had come in except number 12. Lightoller was nursing it in, with 75 people aboard. (Everyone on his rickety collapsible B craft had been transferred to number 12.)

8:30 AM -- Number 12 was alongside. Colonel Gracie said he felt like kissing the deck of the *Carpathia* as he boarded. Harold Bride felt himself being lifted up. Then he passed out. When he came to, in someone's stateroom, a woman was bending over him, brushing his hair, rubbing his face.

Before we go on to New York with the *Carpathia* and her quadrupled passenger list, let's turn the clock back just a little, to 4:00 AM, when the *Carpathia* spied the first *Titanic* lifeboat. This is a watch-changing time on all ships, and this was when Chief Officer Stewart of the *Californian* came up to the bridge to relieve Second Mate Stone. He was told strange tales of what had been happening through the night -- mysterious ships and lights, futile attempts with blinker lights, rockets -- ah! one report says that at the word "rockets" he acted like one himself and ran to wake up wireless operator Evans to find out who had been in trouble. This is only one possibility. It's so difficult, unraveling the *Titanic* story, to be able to say with 100% certainty this or that happened. So much is secondhand, thirdhand, remote, unconfirmed or contradicted. The ship's log of the *Californian* was so sloppily kept, the word "rockets" never appears! One of the best investigative reporters has been Walter Lord, with his two *Titanic* books. He says that if he could have a magical time machine to whirl him back to a certain definite point and place in history, he'd choose to be in Captain Lord's cabin when he was conferring with his Chief Mate Stewart about the early morning watch of April 15, 1912. There is more than a suggestion of coverup, lies, cowardly evasion. Walter Lord is the spearhead* of the attack on Lord, which was begun by William Smith at the U.S. Inquiry and continued by Lord Mersey at the British one. He was never charged with anything, and he was never given a chance to clear himself officially, though he begged for this repeatedly. Briefly, he had a very rough deal, as far as democratic justice goes, in both the U.K. and the U.S.A.

The mystery about Captain Lord's conversation with his chief mate, and about that time from 5 AM to 5:30 AM, is what intrigues Walter Lord. All we know for sure is that Evans was awakened at 5:30 AM, and asked other ships about just what had been happening during the night.

The *Mount Temple* and the *Frankfurt* told him the news: the world's largest ship had gone down, with heavy loss of life, about 20 miles away--and she'd started sending SOSs only minutes after he'd gone to bed! He told the chief mate, who told the captain. Captain Lord came into the radio shack and got it confirmed by Evans.

---

*Walter Lord objects to being called this, but Captain Lord tried to get a fresh trial on the basis of the ill-fame he got when "Night to Remember" was printed, 1955.

He has been attacked by many, defended by some, for not going to the assistance of the *Titanic*.

Attackers say *: He didn't respond as he should have done to obvious signals of distress. He was overly concerned with the color of the signals. He was overly concerned with the lack of any response by the unknown ship to his deck officer's light signals. He didn't do the obvious thing by present-day standards -- awaken the wireless operator. Also (this is implied), he was anchored for the night, terrified by icebergs all around him, and would have used any flimsy excuse or uncertainty to avoid going back into that ice-covered area.*

Defenders say *: He was being realistic about the ambiguity of the rockets (they could have been company signals). Because of the failure of his deck officers to get a response by light signals, he was justified in not doing anything until he was quite sure. Also, he was very, very tired (18 hours on the bridge) and was making the errors of judgment that fatigue causes. About this fatigue there was no doubt; he had fallen asleep as the cadet was talking to him, and he could remember very little of what happened during the night.*

STILL, MY CONTENTION IS THAT LORD'S GUILT OR INNOCENCE IS *IRRELEVANT!* EVEN IF HE HAD RAISED ANCHOR AND HEADED FOR THE *TITANIC* IMMEDIATELY, THE *CALIFORNIAN* COULD NOT HAVE SAVED THOSE ON THE *TITANIC*.

Captain Lord's fears about surrounding icebergs were not imaginary. See the picture which describes the enormous ice field on Sunday, April 14, 1912 -- *exactly as reconstructed by the U.S. Hydrographic Office from reports sent in by ships.* The positions of the bergs were more or less the same all day. At 12:45 AM, the *Titanic* sent up her first rocket. The *Californian* is where shown. The *Carpathia* had received an SOS from the *Titanic* 10 minutes before, and was headed northwest at full speed.

Captain Rostron of the *Carpathia*, who saved 712 people, always will be the hero of the day. He got the SOS at 12:05 AM from 58 miles to the southeast of the *Titanic*. He immediately went to the rescue at full speed. He didn't see any icebergs for two hours and ten minutes. From 2:45 AM, they began to appear, and then were very frequent. Captain Rostron said, "When day broke and I saw the ice I had steamed through during the night, I shuddered, and I could only think that some other Hand than mine was on the helm during the night." One is inclined to agree with him, and to add that Lady Luck was on his side right from the beginning. If he'd been in the position of Captain Lord of the *Californian* and had made a bee-line for the *Titanic* at full speed, he'd have surely hit an iceberg and sunk before he ever got near the *Titanic*.

Lord Mersey at the British Inquiry summed up his views on the *Californian* (and Captain Lord) by saying, "When she first saw the rockets, the *Californian* could have pushed through the ice to the open water without serious risk, and so have come to the assistance of the *Titanic*. Had she done so, she might have saved many, if not all the lives that were lost." This statement was accepted, unquestioningly, for 75 years. It's just not true.

Let's move forward in time to four days after the disaster -- April 19 -- when Captain Lord was giving to the press in Boston *his* account of what happened. He said the first news about the *Titanic* had been received from the Allan liner *Virginian* at 5:30 AM. The *Frankfurt* and the *Mount Temple* had confirmed this. The *Virginian* had given the *Titanic* position as 41.46N, 50.14W. The *Californian*'s position at that time (where, therefore, she'd been all night) was 42.5N, 50.7W. These figures were logged -- no doubt about them. They fix the distance between the *Titanic* and *Californian* at about 19.2 miles. One has to add "about" because minor errors, due to faulty navigation or incorrect clocks/chronometers, are always likely. Even so, we can be sure that the *Californian* started heading for the *Titanic* at about 5:45 AM on June 15, and had approximately 20 miles to go. With this starting point and starting position in mind, let's continue with Captain Lord's statement to the Boston newspapers:

"We started for the scene as quickly as possible. At best, however, it was slow going. At times, nervous and anxious as we were, we hardly seemed to be moving. We had to dodge the big bergs, skirt the massed field ice and plow through the line of least resistance. For three full hours, we turned, twisted, doubled on our course -- in short, maneuvered one way or another -- through the winding channels of ice."

There's one slight exaggeration in that statement -- "three full hours." The actual time taken was two and three-quarter hours. Lord arrived on the scene -- where the *Carpathia* already was -- at

8:30 AM. Otherwise, we can accept his version of the horrifying, iceberg-dominated voyage. A glance at the chart compiled by the U.S. Hydrographic Office confirms the difference between the voyages of the *Carpathia* and the *Californian.* Even Captain Rostron "shuddered" when daylight showed what he'd just been through. Rostron had started his journey in clear, untroubled waters -- going for two hours before he saw a berg. Lord had been immobilized by bergs before he dared raise his anchor.

Now, here's a vital point in the whole story: Just when did the *Californian* get what might (should?) have been interpreted as a signal of distress? When were the rockets first seen? Instead of saying 12:45 AM, let's call it "quarter to one" Then, because we know the *Titanic* sank at 2:20 AM, we can see vividly that the *Californian* had exactly one hour and 35 minutes to get to the *Titanic* before she disappeared ("forever" was to have been the next word, but instead...) until September, 1985. If she could (and did) make that precise journey in daylight, taking two and three-quarters hours, how could she possibly do the same journey in one hour and 35 minutes in darkness, with icebergs looming up every few seconds? Let's suppose there wasn't any darkness -- that everything was seen in broad daylight. Let's suppose that the *Californian* could and did go at the greatest speed she could, just as she did in the daylight. She wouldn't have gotten to the scene until 3:30 AM, one hour and 10 minutes after the *Titanic* had sunk. Then stop supposing. It was dark and the ship had to go slower, much slower than she would in daylight. The *Californian* couldn't possibly have reached the *Titanic* until 5 AM or later.

So there we have it. Even if the *Californian* had responded instantly (i.e., when the first rocket went up at 12:45 AM), one glance (let alone a careful study) at the hydrographic map and the iceberg situation, plus a swift calculation involving the *Californian*'s position, speed, movement, etc., show that for her to have saved anyone on the *Titanic* was impossible. MGY took two hours, 40 minutes to sink. The *Californian* would have taken two hours, 45 minutes if she'd gotten the first SOS at 12:05 AM (which she didn't and couldn't) and had immediately gone at *daylight speed* to MGY position. She'd have found about 1,000 people in the water, and some alive.

But that wasn't the way it was; couldn't be. The first SOS wasn't heard. Even if it had been, the *Californian* couldn't have gone at daylight speed and have arrived at MGY without hitting an iceberg herself. Comparisons with the *Carpathia* are not just invidious, they are remote and impossible.

Perhaps Captain Lord was just tired and stupid. Perhaps he was claustrophobically terrified by those surrounding icebergs. These or any other "perhapses" are irrelevant. They're completely irrelevant. HE COULDN'T HAVE MADE IT TO THE *TITANIC* IN TIME.

Rostron made arrangements for the *Californian* to continue the search for survivors, who were carefully counted several times -- 712. Then as many as possible of the *Titanic* boats were hauled aboard -- six on the forward deck, seven on the *Carpathia*'s own davits. The others were set adrift.

Before he left the scene, Rostron held a burial service for those lost. The *Carpathia* hurried on to New York.

Monday, Tuesday and Wednesday, April 16 to 18 (on board the *Carpathia*). The radio gear of the *Carpathia* was crude, even by 1912 standards. It had a range of about 150 miles. The whole world wanted information about the disaster: the survivors, those who had been lost, and the names of all. The families of 2,229 people, on both sides of the Atlantic, wanted to know if their loved ones were safe or dead. Newspapers, radio stations and ships were all avid for information. All these people and all these organizations directed their attention to the radio link on board the *Carpathia*. Young Cottam bore the brunt of them all. Every station on land or sea was calling him. For three days and nights, he was the sole link between 712 survivors and their families. Captain Rostron forbade him to send anything but official and passengers' messages. Cottam had no sleep or rest through Sunday, Monday and late Tuesday, when he collapsed over the key. There was another wireless operator aboard, the *Titanic*'s Harold Bride. His feet had been frozen in the icy water on the raft, and he couldn't stand or walk. However, when Cottam collapsed, they propped Bride up in a chair, and he continued sending the list of survivors to the USS *Chester*, which had a more powerful transmitter to contact shore stations.

There were some petty difficulties concerning a message the USS *Chester* had for the *Carpathia* from President Taft. (He'd had a personal friend on the *Titanic*.) The Marconi Company was boycotting messages from ships not using their apparatus, and the *Chester* carried De Forest gear. Also, a message from Bruce Ismay aboard the *Carpathia* was delayed. Cottam did not send it for two days, because it was "not customary to put official news through any station at all but those of the same shipping line." He was a little confused on this point, but there undeniably was divided authority in those early days of radio, the Marconi Company versus the shipping line.

Petty trade wars like this were going to be obliterated by the enormity of the *Titanic* disaster.

Wednesday, 18th April -- Waiting for the *Carpathia* to arrive, the news-hungry American and British nations needed someone to blame and someone to praise. They had already been provided with one scapegoat: Bruce Ismay (owner of the *Titanic*) was guilty of being alive. Scores of women and children had died, but he had managed to get saved. The other scapegoat was to be Captain Stanley Lord of the *Californian*.

Captain Rostron was the ready-made hero. He had driven his ship at full speed, skirting dangerous icebergs and pack ice. He had known exactly what to do throughout, and had done it. Later, President Taft, for the Senate, gave him a Congressional Medal of Honor. The *Titanic* survivors gave him a loving cup, handed over by "Unsinkable Molly Brown."

All the Marconi operators using Marconi gear brought their originator into the limelight. It didn't do his stock on the New York Stock Exchange any harm. In two days, it went from 55 to 255.

Colonel Gracie, one of the passengers, whose account of the *Titanic*'s end is considered one of the best, spoke for all his shipmates by saying, "In the midst of our thankfulness for deliverance, the one name mentioned with deepest feeling of gratitude was that of Marconi. I wish that he had been there to hear the chorus of gratitude that went out to him for the wonderful invention that spared us many hours and perhaps days of wandering about the sea in hunger and storm and cold."

Marconi himself said that the one lesson to be drawn from the *Titanic*'s loss "is the necessity or at least the desirability of having two wirelessmen on every ship." This would have dealt with the tragic irony of the nearby *Californian*, but the basic idea of continuous watch on 500 khz (by auto-alarm and/or human watch) was still in the embryonic stage.

The welcoming ceremonies prepared for the *Titanic*'s arrival in New York had to be readjusted for the incoming *Carpathia*. The fleet of tugs had no banners; they did not joyfully hoot their horns. The gay decorations all around Pier 54 were quietly removed. Pier passes were issued to relatives, police, nurses, doctors, city and government people. Twenty ambulances stood by. The railroads were offering free transportation to all immigrants, and relief workers were ready to help the indigent and homeless.

Marconi officials boarded as soon as they could, and went in search of Harold Bride. He was still in the radio shack, with a hundred unsent messages in front of him -- eyes sunken, face colorless. He didn't know the *Carpathia* had docked, that his work was over. Stretcher-bearers took him ashore.

Then the Marconi people looked for Cottam, the *Carpathia* operator proper. After the three most harrowing days of his life, he was in a deep sleep. He'd had 18 hours of it so far. They crept away and let him take his rest. Let us, too, leave him there recuperating, while we go forward a little in time, searching for missing bodies. After that, we'll return to the sleeping Cottam. He, we, everyone, is going to need restful composure to cope with William Alden Smith, the furious senator from Michigan, who'll spare nobody, least of all himself, until he's got to the bottom of the *Titanic* story.

There's another reason to return to and reflect on Harold Cottam (lying utterly exhausted as the *Carpathia* pulls into New York). We'll leap far into the future, then come right back to him.

On the 70th anniversary of the Titanic *disaster, in 1982, the Society of Wireless Pioneers will make a lot of little* Titanic *pins and will assign yours truly a double task: (1) verify or disprove the rumor that Harold Cottam is still alive; (2) if he's alive, present him with a Titanic pin. A search of birth/death records showed that Cottam was alive -- somewhere in England; but both the British government and the* Titanic *Historical Society were protecting him and all other* Titanic *survivors; the only way to get the pin to him was via the Department of Pensions. This was done. Two years later, Cottam died, aged 94, and his story was in newspapers everywhere. It was then realized, far too late, that he never got the credit he richly deserved.*

Back to 1912. You've already had a preview of Senator Smith, interviewing Fourth Officer Pitman. Smith was, as President Taft said, "a bobbing sort of cat, who wouldn't be sat on." No ... wait, Senator. We want to know what happened to all the *Titanic* corpses -- mostly frozen, some drowned.

The bodies picked up by the lifeboats were few. Most of them were caught up in the immense ice-pack going northeast with the Gulf Stream. The lifeboats, and then the searching ships, had kept away from the ice, so they didn't see them. When the ice broke up, bodies were seen. One was 25 miles from the original point. One ship, SS *Bremen*, saw 150 bodies, including a man in evening dress on a door -- two more on steamer chairs -- and a woman in her night dress.

The *Mackay-Bennett,* chartered by White Star Line, came back eight days after the sinking, as a "morgue ship." Icebergs and "growlers" (overturned bergs) were everywhere. So were hundreds of bodies. All in life jackets. All dead from cold. Some had crushed skulls but were identifiable. These were embalmed. The unidentifiable ones were buried at sea. Add heavy weights to them. Wrap 'em in cold canvas. Over the side. Splash. Splash. Splash. Down two miles to join the *Titanic*. Altogether, the *Mackay-Bennett* found 306 bodies; buried 116 of them at sea and brought the rest to Halifax. People came from all over the U.S.A. and Britain to the curling rink where the bodies -- some decomposing -- lay. One of the corpses was multimillionaire John Jacob Astor -- crushed to a pulp and only recognizable by his large diamond ring.

Where were the rest of the bodies? 1517 minus 306, equals 1211 -- all missing, never found. At 6,000 pounds pressure per inch, some would be compressed into jelly messes. Others would be part of the Gulf Stream and flow around with it -- eastward to the British Isles--then some going around Scotland to the North Sea -- others turning south past Portugal -- southwest by Morocco -- then westward back to where they came from. Some, of course, would be washed ashore. Some gently gesticulating skeletons must have gone round and round the Atlantic for many months -- maybe years, maybe *still*, NOW!

There was a bit of a scandal around Bride a few days later. It concerned some messages he'd received just as they were docking, and just before he collapsed at the radio console. The U.S. Navy had intercepted the messages from the Marconi Company to both Cottam and Harold Bride, on the *Carpathia*: SAY OLD MAN, MARCONI CO TAKING GOOD CARE OF YOU. KEEP YOUR MOUTH SHUT AND HOLD YOUR STORY. IT IS FIXED FOR YOU SO YOU WILL GET BIG MONEY.

Then, later -- ARRANGED FOR YOUR EXCLUSIVE STORY FOR DOLLARS IN FOUR FIGURES, MARCONI AGREEING. SAY NOTHING UNTIL YOU SEE ME.

This referred to Marconi's exclusive story (from Harold Bride) for the *New York Times.* The Marconi-*Times* collusion sparked off indignant headlines by other newspapers -- TITANIC STORY IS HELD UP FOR CASH -- NEW YORK PAPER KEPT WORLD IN AGONY WHILE BICKERING FOR NEWS.

This was not true. The *New York Times* offer did not come until the *Carpathia* was docked in New York. With the *Titanic* story, Editor William Van Alda steered the *New York Times* from being just another New York newspaper to its eventual encomium as "the world's greatest newspaper." Marconi said he had no objection to the wireless operators receiving monetary reward for their heroism during the disaster. Senator Smith, as a keen-witted outsider, had a simpler solution: "Why not just pay the operators a decent wage?" ($20 a month didn't go far, even in those days.)

The American Inquiry began on Friday, April 19, Michigan Senator William Alden Smith presiding. He mercilessly questioned everybody and everything and issued subpoenas for British citizens.

Smith arrived in New York just about the time the *Carpathia* was steaming into the harbor. Determined, if necessary, to subpoena everyone on board to a congressional hearing, he dashed up the gangplank, identified himself to the police, and asked for Ismay's cabin. On the door was a notice, "Please Do Not Knock." Smith knocked, of course, and then forced his way in. Within an hour, he was telling reporters that he'd had a frank, courteous interview with Ismay, who would be with the four surviving officers of the *Titanic* at the Waldorf-Astoria the next day to see the subcommittee.

THE AMERICAN INQUIRY

The public, along with umpteen reporters and photographers, were all present to see Senator Smith attack, probe, doubt and question Ismay. The White Star boss said he'd helped put women and children aboard the lifeboats, then left on a collapsible boat--the last boat to leave the ship, as far as he knew. He said the number in his boat was 45. There were no passengers on deck when he left. The boats were the correct number required by the British Board of Trade. Smith softened his assault on Ismay toward the end, especially because he had found a much more likely culprit, Captain Lord. Ismay's real culpability for the *Titanic* disaster had occurred a year before the ship was launched -- and it wasn't discovered until 70 years later. Senator Smith took some of the heat off Ismay, but the British Inquiry disgraced him, arguing that it had been his "duty" to go down with the ship. He gave up his official posts as president of International Mercantile Marine and White Star chairman. He left the public eye completely, retired to the west of Ireland, and was seldom heard from again. He did, however, contribute to the Mercantile Marine Widows' Fund, thus retroactively helping the Titanic widows. He died October 17, 1937.

Friday, the 19th, the *Californian* docked in Boston. Her Master, Stanley Lord, queried by Associated Press, said:

"About 10:30 that Sunday night, we steamed into an immense ice field and immediately, as a matter of safety, our engines were shut down to wait for daylight. With the engines stopped, the wireless (operator) was not working, of course, so we heard nothing of the Titanic's plight until next morning."

To the *New York World* he added:

"Up to the moment of shutting down, no message of distress or any signal was received or sighted. The first thing the *Californian* got was a confused message from the *Frankfurt*, from which we finally made out that the *Titanic* was in distress.... We were from 17 to 19 miles distance from the *Titanic*, quite outside normal visual range."

(If so, how come lights and eight rockets were clearly seen by others on the *Californian*? The distance could have been exaggerated -- 10 to 15 miles became 17 to 19 miles -- but the lights and rockets were not to be dismissed so easily.) Third Assistant Engineer Ernest Gill was irate when he read his captain's account. He and four fellow engineers swore out an affidavit, to appear in the *Boston American:*

"On the night of April 14th, I was on duty in the engine room from 8 PM to midnight. At 12:30 AM, I came on deck. The stars were shining brightly. It was very clear. I could see for a long distance. I saw a large steamer about 10 miles away. They could not have helped but see her from the bridge and lookout. I turned in but could not sleep. In half an hour, I went out to smoke a cigarette. After 10 minutes, I saw a white rocket about 10 miles away on the starboard side. In seven or eight minutes, I distinctly saw a second rocket, same place, and I thought, That must be a vessel or lookouts. But they could not have helped but see them. I knew no more until I was awakened at 6:40 by the Chief Engineer, who said, 'Turn out to render assistance. The *Titanic* has gone down.' Then I heard the second and fourth engineers in conversation. The second said the third mate had reported rockets had gone up on his watch. The captain had been notified of the rockets by the Apprentice Gibson. Mr. Stone, second mate, was on the bridge at that time. More lights were seen. More rockets. Mr. Gibson went to the captain again. The skipper told him to continue to Morse until he got a reply. Then I heard Mr. Evans (second engineer) say, 'Why in the devil didn't they wake the wireless man up?'"

Ernest Gill's statement had a touchingly sincere ending:

"I have no ill will toward the captain or any officer of the ship, and I am losing a profitable berth by making this statement. I am actuated by the desire that no captain who refuses or neglects to give aid to a vessel in distress should be able to hush up the men."

The *Titanic* sent up eight rockets. The *Californian* officers-on-watch saw eight rockets. Captain Lord was informed about at least four of them. Regarding the others, he may have been half asleep or asleep.

"You never mistake a distress rocket," said Captain Lord himself at the inquiry! Was there room for doubt, as Lord stated, owing to confusion between low-lying stars, shooting stars, company signals and low trajectory?

Lord was very prudent, ultra-cautious. Captain Rostron would have reacted and hustled round to his wireless operator's cabin immediately.

Lord said nothing to the press about rockets, and there was nothing about rockets in the *Californian* log book!

He told Smith that "between 1:15 AM and 6:30 AM I have a faint recollection of the apprentice opening my room door."

Captain Lord did not, in his opinion, have enough incontrovertible evidence that there was a ship in trouble very near him. The officers on watch and crew members saw rockets, but there were other uses for rockets, as company signals. Even so, how could physical tiredness befog his mind so much that he wouldn't react to rockets at all?

Smith's summary of the "*Californian* situation" (as it was to be slightingly called) was: "Why did the *Californian* display its Morse signal lamp continuously for nearly two hours if they saw nothing? The signals which were visible to Mr. Gill at 12:30 and afterward, and which were also seen by the captain and officers of the watch, should have excited more solicitude than what was displayed by the officers of that vessel. The failure of Captain Lord to arouse the wireless operator of his ship places a tremendous responsibility upon this officer, from which it will be very difficult for him to escape. Had he been as vigilant in the movement of his vessel as he was in displaying his own signal lamp, there is a very strong possibility that every human life that was sacrificed through this disaster would have been saved."

The British Inquiry castigated Lord thoroughly. He received the stigma of a British master who had not gone to a vessel in distress. His license was not revoked, but his resignation was requested. He kept trying, unsuccessfully, to have his case reopened, and he retired from the sea in 1927. In 1958, there was a renaissance of interest in the *Titanic* (when Walter Lord made indirect attacks on Captain Lord, who again pressed to reopen his case and prove his innocence.) Walter Lord's almost direct attacks were made without knowledge of the defense outlined in this book. They were far too late to help Captain Stanley Lord. He died in January, 1962, age 84.

Senator Smith contrasted Lord's conduct with that of Rostron, who "by his promptness and indifference to peril" charged across 58 miles through icebergs to the Titanic's position, and saved 712 lives. Smith proposed a resolution calling for a $1,000 medal to be presented to Captain Arthur Henry Rostron by the president of the United States. This resolution was approved by acclamation.

Rostron commanded the Cunard's flagship *Mauretania* from 1915 to 1921. He retired in 1931, wrote his memoirs, and died November 4, 1940.

Second Officer Lightoller of the *Titanic* (38, 13 years at sea) evaded some of Smith's questions, but was on the whole forthright. He was vague about his knowledge of icebergs. He had been in charge of boats on the portside. He had been uneasy about the huge drop (70 feet) for the boats. As the *Titanic* went down, Lightoller had dived from the roof of the wheelhouse. He was being drawn, by suction, to the officers' quarters in front of the funnel. Then hot air from a boiler explosion had set him free. After going down and up twice, he found himself near collapsible boat B. He defended the White Star line in the British Inquiry. Many years later, in 1940, he commanded his family's 60-foot yacht at Dunkirk, and rescued 130 men. He died in 1952, at 78.

Cottam, when called to the stand, looked a mere boy, though 23. He impressed everyone by his account of his awareness of his grave responsibilities -- with unregulated watch hours, perhaps continuous, for $20 a month. (He was the unsung hero -- see my remarks later.)

Boxhall, the *Titanic*'s sixth officer, stayed with the White Star Line until he became first officer of the *Aquitania* in the 1930s. He died in 1967, age 83, and his remains were scattered (at his request) over latitude 41.46N longitude 50.14W.

THE SMITH BILL

Senate Bill No. 6976 radically altered marine legislation. Future passenger-ship construction had to include longitudinal bulkheads as well as transverse ones. There had to be enough lifeboats for all on board, which had to be lowered, at least in practice, twice a month. The proposals about wireless were most stringent, anticipating the S.O.L.A.S. Convention of 1914: (A) No steamer carrying 50 passengers or more would be permitted to leave U.S. ports without a wireless set, having auxiliary power, and a minimum range of 100 miles. (B) There had to be direct communication from radio operator to bridge and two or more wireless operators for continuous watch. Any master not enforcing these regulations was to be heavily penalized. Rockets were to be sent up *only* for distress. (This would remove any gray uncertainty that people like Lord might have as an excuse.)

William Alden Smith went on supporting unpopular causes -- *for* black Americans -- *against* U.S. entry into World War I -- *for* Hoover, *against* Roosevelt in 1932. After a passionate speech for Hoover, he died of a heart attack. All but forgotten now is the good he did by the Smith Bill to all seafarers.

"WOMEN AND CHILDREN FIRST"

In England, Mrs. Chapman, arch-suffragette, said she'd a thousand times rather go down with the ship under similar circumstances. Her American counterpart, Harriet Blatch, said, "Since men had drafted the laws about the ship, they should have been the ones to go down with it." Women, she intimated, would have insisted on laws requiring plenty of lifeboats.

The "women and children first" call aboard the *Titanic* inspired antisuffragettes to raise a monument in the East Potomac Park in Washington, subscribed to by single dollars from 25,000 women. This 18-foot statue is a half-clad male, arms outstretched. On the pedestal is engraved, "To the brave men of the *Titanic*, who gave their lives that women and children might be saved." Here are the figures:

	Women and Children Saved	Men Saved	Total Saved
1st Class	96%	3%	60%
2nd Class	81%	10%	46%
Steerage	47%	14%	25%
Crew (2 men per lifeboat)	87%	22%	26%

CLOSING WORDS ON THE TITANIC

(WHAT THINGS/EVENTS/PEOPLE MIGHT HAVE PREVENTED THE DISASTER ... IF ... IF ... ETC. ???)

HW = Hindsight wisdom

WTS = Wrong time slot

The IFs	Related to...	Whose Fault?
1	Insufficient lifeboats.	Many. But all passenger ships in those days had too few. WTS
2	Speeding in dangerous area, careless about warnings, etc.	Traditionally, and rightfully, captain's responsibility. But he was *typical*. WTS/HW
3	Whole attitude of everybody (especially navigating officers).	Again, typical. WTS/HW
4	Earlier sighting of iceberg -- even by a few seconds.	Who omitted binoculars? Or was it accident, to be expected on maiden voyage? HW
5	Later sighting (perhaps from inattention/ sleepiness/whatever) would have been a blessing in disguise.	No way this could have been foreseen. HW

6	24-hour radio watch.	WTS/HW
7	Auto-alarm	WTS invention.
8	Captain Lord and *Californian*.	THE scapegoat from the beginning. About him personally -- no comment. His intentions/character irrelevant. He COULDN'T (even if he would) have rescued the survivors.
9	Cyril Evans (Sparks, *Californian*).	He wasn't required to be on watch. He wasn't called. WTS
10	Iceberg messages the *Titanic* navigators did NOT receive.	Phillips and Bride did pretty well. The missed messages were not really new, vital info. Navigating officers had already been alerted -- and they weren't alarmed enough! (See 3 above)
11	Murdoch should have signaled FULL SPEED ASTERN and nothing else.	HW -- and almost impossible to fault Murdoch (though his suicide, if true, was probably from self-blame).

12	*Californian* receiver not working when third mate listened (just as SOS being sent).	WTS. This has been judged the most poignant IF.... IF Evans heard the SOS, Captain Lord would have had no excuse for not proceeding immediately to the *Titanic*. Even so (as pointed out previously), there couldn't have been a rendezvous until after 3 a.m. (See page 320.)

All these IFs hinge on...

Change of circumstance(s)

The clash between prevailing conditions and what would/should have been, IF...

Mostly, the unfortunate timeslot the Titanic was in -- plus everyone and everything connected with her.

There is no real BLAME attached to *ANYONE*!

THE GOOD THAT CAME OUT OF THE TITANIC HORROR

Immediately -- Sailing routes, all ships were shifted south of the iceberg region.

Almost at once -- Lifeboats for all became a general requirement; this was made officially mandatory by...

January, 1914 -- when the first S.O.L.A.S. met, and laid down other international maritime laws.

The International Ice Patrol was an outcome of S.O.L.A.S. It had been de facto almost immediately, managed by British, then American ships. When it was officially born, all nations supplied money and the U.S. Coast Guard (in Woods Hole, Massachusetts) was put in charge. They have run it ever since. *No lives have been lost in the patrolled area from that day to this.* Important outgrowths of the patrol's work were: Sonar (soundwaves in water to detect icebergs, etc.) and fathometers (for water depth). The USCG still drops a wreath every year over what used to be regarded as the approximate position of the *Titanic*; it's exact now, thanks to Dr. Robert Ballard and to U.S./French governments' money. The Ice Patrol (now doing most of its work by plane) is the longest running and most successful international effort ever.

The New York Times had its greatest turning point, by its masterly coverage of the disaster. Under Editor Carl Van Alda, it went on to become the world's greatest newspaper.

Radio (wireless) had its greatest turning point marked by the *Titanic* and S.O.L.A.S. Spurred by the ironic tragedy of the *Californian* being only 19 miles away, with her radio closed down for the night -- S.O.L.A.S. decreed 24-hour monitoring of 500 khz (international calling and distress frequency). At first this was done by human watch; in 1928, auto-alarms awakened the sleeping R/Os. There were many radio regulations laid down by successive S.O.L.A.S. meetings.

A SAD POSTSCRIPT ON HAROLD COTTAM

Harold Cottam's importance as the Sparks on the *Carpathia* cannot be overestimated. Without him there probably wouldn't have been any survivors. They couldn't have lasted more than a few hours in the boats; those in the icy water were doomed anyway.

In response to the *Titanic* SOS, he stuck to his post for 65 hours (only seven hours less than three full days), relaying survivors' names, with Harold Bride of the *Titanic* trying to assist him for a short while. Cottam has been grossly underestimated--at the time and ever since. Captain Rostron got headlines, glory, reward, money--but who told him about the *Titanic*? Who pressed the button to start everything?

Cottam's name wasn't even mentioned in the *New York Times*, nor in any other newspaper! He then faded out of notice completely for 73 years. Resolute efforts to locate him got nowhere, as mentioned: *Titanic* survivors were over-protected by the British government. On his deathbed, age 93, he was remembered, but only because anyone/anything to do with the *Titanic* had/has news value. He died 15 months before the *Titanic* was found. An informal obituary by someone who knew him said, "He was a cheerful old codger, lively and well-liked." That's nice to know, but Cottam should have been recognized as one of the world's heroes.

FIRST DOMINANT & LAST } WORD ON THE TITANIC

Before you had read these viewpoints, theories, etc., all massed behind that stark, trisyllabic TI-TAN-IC, it's very likely that you had read some other accounts or stories. Such complications! So many "maybe's." All those IFs. Food for thought everywhere! With such interleavings and complications, it's hard to know where to start sorting everything out.

This was also a problem for super-*Titanic*-buff Walter Lord when, 30 years after his famed *Night To Remember*, he reapproached the subject with *The Night Lives On.* He solved the dilemma by constructing a "basic framework of the calamity" from 11:40 PM on April 14, 1912, to 2:20 AM on April 15. This method has been borrowed from him, with thanks.

However, that's only begging the question, really. To select a device (which is all a clock is) and measure everything by it gives us only a semblance of order and rationality. Some things will forever remain unexplained and unexplainable.

Before we get around to what should have been regarded as the central reality for the whole *Titanic* story, let's deal parenthetically with something that should have been, especially in the minds of radiomen of all kinds, the most important thing to do: Why the heck didn't they wake up the radio operator, Evans? Well, note his meager title. "Officer" he was not. He got a salary of $20 a month. Doesn't that show how low-esteemed he was...how unimportant he and his contraptions were? Oh, there had been one spectacular demonstration of the usefulness of radio (the *Republic/Florida* incident in 1910), but the lesson wouldn't be really learned until mankind was recovering from the *Titanic* disaster. What is simple and obvious to us now wasn't so in 1912.

In fact, there was one overwhelming REALITY. One word: ICEBERGS. ICEBERGS. ICEBERGS. Everything else devolves from it or comes back to it.

Just look at that map from the U.S. Hydrographic Office, showing where all the icebergs were on April 14. Stare hard at it. See where all the ships were. Remember that those little marks showing the *tips* of icebergs only indicate the vast, thick *field* of ice beneath the surface. Then, fully ice-conscious, consider again the situation of the *Titanic*. There are thousands of details -- many very interesting--but all distracting you away from that crude, jagged reality.

Look at the map again. It only shows ships, already *in* that area. Immediately after the *Titanic* sank, that part of the Atlantic became a big no-no, and stayed that way. Let us finally recognize that, dominating everything there, were ICEBERGS, ICEBERGS, ICEBERGS.

Icebergs, to brave, jolly old Captain Smith, were mere incidentals in the background of navigational details and his social life as the Commodore of the White Star fleet while he triumphantly merged this maiden voyage with his last trip before retiring for good.

Icebergs, to ultramethodical Captain Rostron, were important things to cope with as he dashed, top-speed, to the aid of the *Titanic*. Later, when he saw in broad daylight what he'd just been through, he blanched. Yes, the hero of the whole situation said, "I shuddered...some other Hand than mine was on the helm during the night."

Icebergs, to cowardy-custard Captain Lord, were a horrible, all-embracing fact, and he anchored for the night. To dash off, helping someone who might be in trouble -- in the dark, with all those icebergs around -- was suicide. (It really was. Look at that map again.) If he could see, without lookouts at all points, and go very carefully, as he actually did in daylight, taking three hours to get to the Titanic -- ah, that's a different matter.

Walter Lord (amazing, that coincidental duplication of his and Captain Lord's names) wants a time-machine to whirl him back so he can listen in on Captain Lord and Chief Mate Stewart chatting at about 5:00 AM, just before awakening their Sparks, Evans. I don't. The details would be interesting, but I have no doubt at all about the main topic: ICEBERGS!

The enormouse ice field -- revealed by daylight to horrified Captain Rostron -- was a huge barrier of "growlers", loose ice and icebergs of every kind and size. This was the feature of the *Titanic* disaster beside which all else was relatively insignificant.

QTP
means

"I am
entering
port."

A Quiet QTP

Our engines just revving.
(DEAD SLOW)
we pass a Swedish ship
then a British one
and drop anchor.

Two cormorants flap noisily down
and settle in the water
together -- always together --
they copy, copy.

Here comes a motorboat
Chuggachuggachuggachug ... it stops
by the rope ladder, portside;
the agent climbs aboard.

I'm watching it all
from the flying bridge.
It's so quiet up here
after the cacophony of 500 kiloherz.

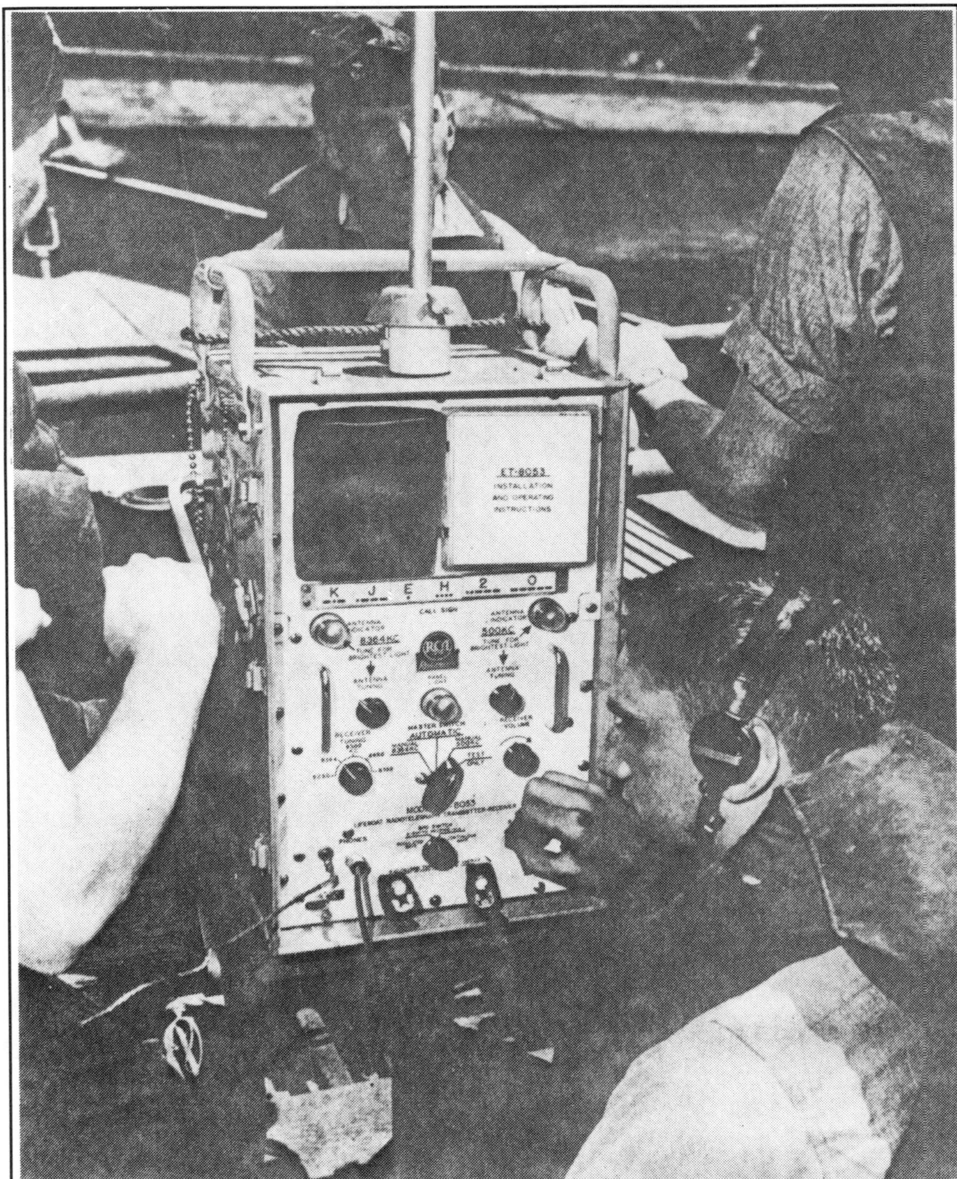

Radio installation in lifeboat

Two sailors (right & left) are turning handles to operate dynamo and provide electric power. Sparks (below) is listening to receiver and keying the transmitter. The upright rod is the antenna; the ground is contacted by having a wire trailing in the sea. The whole outfit can be operated by anyone, even if they don't know Morse; it will automatically send out the 4-second-long dashes to sound auto-alarms on all ships — then they can use direction-finding to locate the lifeboat.

UNION INTERNATIONALE DES TELECOMMUNICATIONS
INTERNATIONAL TELECOMMUNICATION UNION
UNIÓN INTERNACIONAL DE TELECOMUNICACIONES

INDICATIFS D'APPEL
service mobile maritime

CALL SIGNS
maritime mobile service

DISTINTIVOS DE LLAMADA
servicio móvil marítimo

ПОЗЫВНЫЕ СИГНАЛЫ
морская подвижная служба

呼 号
水上行动业务

French, English, Spanish, Russian, Chinese, in that order. It's a throwback to the days when French used to be considered the international language, in politics anyway. All sailors know (Gen. DeGaulle rolls in his grave!) it just ain't so. English is the international language, followed by Spanish.

Cupid's Arrows

versus

Hammer & Sickle

I've been wondering where and how to start this story. Yours truly enters it only toward the end, and I'm of no importance in the story itself. On the other hand, without me you wouldn't know any of it! It's all quite true, with romantic music in the background to make it into a soap opera -- no, more like an old-style melodrama with a Hero, Heroine, Villain and a Knight in Shining Armor to the rescue. I'll start by introducing our Hero.

His name is Tyrone Patrick. Begorrah, 'tis hard to think of a more Irish name than that! However, before you let your imaginations go flitting away like will-o-the-wisps on an Irish bog, you should know that Tyrone has never seen the Emerald Isle. His grandparents came over from Ireland in the steerage class of a ship. He is Texan born and bred. As an able-bodied seaman (AB) on an American ship, he has been pretty well everywhere else, so he's a world citizen in outlook. We first find him in Leningrad, formerly Petrograd, the former capital of Russia. He's a pleasant young fellow, normally easygoing, but now his patience is going to be tried and his Irish stubbornness will come to the top.

Step forward, Heroine. She's a pretty little Russian brunette -- frolicsome, happy-go-lucky. She's a hairdresser, for those who want the full treatment, or just a barber for those who don't. She's always ready for a laugh with her customers. A native of Leningrad, her name is Luda Vratolivna. She's not a gloomy Russian any more than Tyrone is a stereotype Mick full of blarney.

They met when he came in for a haircut. Since then, they had eaten in restaurants together, been dancing, and had been on trips for tourists. Leningrad is particularly good for that. Tyrone was impressed by the cemetery for the 600,000 Russians -- 10,000 in each of 60 long communal graves -- who died, mostly from malnutrition, during the three-year siege by the Nazis in World War II. Tyrone and Luda had reached such a state of mental attunement that they often found themselves starting the same sentence before they dissolved in laughter. You hardly needed to see them nuzzling each other's faces like puppies to know they were in love. We'll call 'em henceforth what they called each other -- Ty and Lu.

The melodrama's director tells his imaginary movie camera to lift up-up-up to give us a panorama. Here's the overall view. Ty's ship was slow in cargo work -- satisfactory from his point of view-- but it wasn't going to stay long enough for our lovers to get what they wanted -- marriage. The authorities had no direct, immediate resistance to daughters of Mother Russia wedding foreigners. In fact, the first steps were easy: just fork out about $5 and register at the Wedding Palace for a date three months ahead. There lay the snag. The bridegroom-to-be would be leaving the country before then and would have to return in time for the scheduled wedding. During that three months' wait, other obstacles could cause delays, even defeat.

If/when Ty reappeared after three months, all papers had to be in order. There had to be OKs from Uncle Sam, Ma Russia, and the nearest relative of the would-be-weds. The Russian government also wanted Ty to refund the cost of Lu's education. This wasn't too hard; she'd had a normal education through high school, then had gone to a trade school for barbering and beauty-shop techniques.

It was much too chancy for him to rely on getting a ship to work his way back. Timing was of the essence. He'd have to fly back. Expensive, yes, but love is determined, blind, unstoppable, etc. They planned strategies and defenses like generals. When his ship was due to leave, they were like spring-wound clocks -- taut with determination. They gushed forth love and optimism to smother those tiny, wriggling fears that they mightn't be able to outwit that monster, Russian bureaucracy. The biggest difficulty of all, getting Lu out of Russia, they dared not think about too much. No matter how persistent and Machiavellian they might be, the Russian government would have the last say.

As Ty's ship was pulling out, Lu was as close as she could get on the dockside, waving, shouting, blowing kisses. He had work to do on deck, gathering in ropes, going wherever the mate or second mate sent him. Whenever he could, he, too, would be yelling to Lu in the distance. She was wearing a bright red and blue dress, and a faint smudge of those colors with no definite shape was the last thing he saw before he went below to his cabin.

Three months of separation. Everyone knows that our conventional way of measuring time by hours, days and weeks becomes meaningless when love and impatient frustration are involved. They jabbered on the long-distance phone by unreliable cable (this was just before satellites came along) until either their money ran out or the static interference won, or both. Lu used her considerable charm and love-inspired cunning to make friends with the Wedding Palace folk. She got their sympathy and support. If the red tape (Ty had grim jokes about the Red red tape) could be stretched, cut, shortened, whatever, they'd do it for these anxious lovers. At his end, Ty worked at a number of jobs ashore--he daren't risk being on a ship and unable to make its schedule jive with his. He had booked his flight via TWA well in advance, had checked and rechecked every detail. The leaden hours and days had been clumping by this impatient lover.

Now, at last, it was time.

In Houston he boarded the plane which was going to take him to New York via Washington, D.C., and then on to Leningrad. There Lu was ready. The lovers were happily tense. Everything was absolutely . . . uh-oh.

The Villain enters.

The Russian vice-consul in Washington was given his usual breakdown of all the passengers who were flying to the Soviet Union. Tyrone Patrick? Who's he? With no previous information or clearances, a strong negative had to be slammed on anything to do with this passenger. *Nyet, Nyet, NYET!*

That was that. Tyrone was almost forcibly removed from the plane in Washington. He raged. It got him nowhere. He tried wheedling, pleading. No good. Months of preparation, hundreds of dollars had been spent one way or another to make what he'd thought was a solid structure. It turned out to be a house of cards, brought tumbling down by a petty little official with a mere finger movement and a squeaky *"Nyet."*

Fortunately, Ty was not one for silent suffering. Back in Houston, he moaned and wailed to anyone, everyone.

As for Lu, she had never been so low in spirits in her life. She had gone to the airport not just her usual bouncy self, but better. She was blown up as tight as a balloon with happy anticipation. As passenger after passenger debarked and Ty did not appear, her balloon of happiness was leaking. When it was quite clear that Ty was not going to appear, her balloon of optimism became a squashy little mess. Worse was to come. The hitherto fairly reliable Houston-Leningrad phone-link had a day off. Two days off. Ty was ranting, cursing, with an uncooperative black phone-piece in his hand, hour after hour. He did, at least, know what and whom to blame: the vice-consul and the shaky, weather-affected phone link. Poor Lu knew nothing. After getting nowhere with her attempts to phone, all she could do was wait and wait in passive misery. Never had her sunny nature been so pressed with gloom. She felt as though she was in a melodrama, winds howling across the steppes of Siberia, while she peers across the snow for a lover who never comes.

Well, eventually the cable lines cleared and the Ty-Lu give-and-take warmed their misery a few degrees. Lu had done all she could -- she *might* stave off the Wedding Palace schedules -- quite illegally -- IF . . . and this is where the blockage at Ty's end had to be removed.

Let's pan out again, and float upward. The situation is black. If there were to be any hope or light -- how and from whom/where could it come? The Hero and Heroine were in despair. The Villain was triumphant. We should expect to see (if the melodrama is running its proper course) a Knight in Shining Armor come to the rescue.

He appeared, on cue. Mike, a columnist with the *Houston Post*, heard about Ty's problem. (How not? Ty was almost yelling from the rooftops.) Lovers thwarted at last moment . . . Cupid nearly successful but Iron Curtain clamps down . . . Scrooge-like meanie in Washington turns love story into tragedy . . . mmm, yes. Sounded interesting. Columnist Mike questioned Ty, and soon got the whole picture.

In his column he blasted the Russian vice-consul, and copies of it were sent to that gentleman himself. Presto! It was as though a fairy godmother had waved her wand. (Wands can get magical results more swiftly than a well-meaning but plodding Knight in Armor.) Ty and Lu had *hope* again.

Once that *"Nyet"* had been converted into a "Maybe . . . if . . .", Ty started again, and this time played the game the way *they* wanted. He bought his ticket at INFLOT, the Russian airline. He got official OKs for everything. All looked well. Three weeks to the day since he'd been ejected from the Leningrad-bound plane he was in another one, really going there.

Lu was there, of course, and they collided in an ecstatic embrace. She'd successfully finagled things with the wedding people. Within hours, they were Mr. and Mrs. Patrick, flying on to Moscow and beyond, on a package-deal honeymoon trip. They saw all the things they were supposed to see, did all they had to do -- and had no real interest in any of it. They had each other, and to them the background was misty, unimportant.

"I'm so happy, sweetheart, that. . ." Ty said.

"Me, too, Ty!" Lu replied. But he hadn't finished his thought.

"It's like, like . . . you know what keeps coming into my mind? Waves on a seashore! Each wave breaks at the top with foam, then there's another . . . and another . . . and another, each one spilling over with a white foam of happiness."

Even as he spoke, he remembered (per his analogy) that a tide of incoming waves has an undertow. Things could still go wrong. They hadn't got Lu out of Russia, and she wasn't yet clear to come back to the States with him. Uncertainty, fear, discouragement -- were they condemned to live with these forever?

The honeymoon ended. Ty's visa expired. He had to fly back, alone. As he looked down at her from the ramp leading into the aircraft, a jagged black thought occurred--he might never see her again! All these soaring hopes and black despairs had made life too melodramatic. Why couldn't they just be left alone? He took a mental snapshot of her as he entered the aircraft. Her smiling, pretty face was there whenever he closed his eyes. They couldn't take that away from him.

Once back in Houston, he had to face a grim financial reality. He was nearly broke. He'd have to get back to sea as soon as he could. Day after day he went along to the hiring hall for a job. It was a slack period. Day after day, not even a prospect flickered. Soon it was two weeks. His money was running out completely. He'd only been able to afford one phone call to Lu; all it told him was that she was equally uncertain and miserable.

After all the violently kaleidoscopic changes and roller coaster of emotions, this grey, dull, actionless stream of hopeless days was very wearing. Then the clouds broke open: a job on the *Gulf Knight,* bound for the Panama Canal, the West Coast and maybe Japan. He grabbed it.

Once again, for the last time, we pan out, lift up a little from the hurly-burly of day-to-day existence and get a different slant on it all.

You, reader, are now in direct contact with yours truly, the teller of this tale. I was the radio officer of the Gulf Knight. I've just come into the picture. I know nothing about Ty, Lu, Russian weddings, thwarted lovers, hopes going skyward, miserable weeks of waiting, etc. I come in at the end of the story, and have to work my way back, by questioning Ty.

This was my first introduction to the whole story:

Radiogram Via ITT RS79

ITT Mackay — A Division of ITT Telecommunications Corporation

All messages are accepted subject to rates, rules and regulations in the applicable tariffs on file with the Federal Communications Commission/I.T.U.

Prefix: MSG	No: 1
Check: 15/14	
Date: 6/6	

Office of Origin: HOUSTON

Time Filed: 1430 gmt

Charges
Ship
Coast
Landline
Tax U.S.
Misc.
Total

TYRONE PATRICK AB SS GULF KNIGHT

VIA KLC/GALVESTONRADIO

YOUR WIFE LUDA HAS ARRIVED

FROM LENINGRAD

MANAGER/

CEDAR HILL APARTMENTS

* Received ☒ Sent ☐ KLC (Station) 6/6 Time: 1450gmt (GMT) Signed:

*THIS FORM IS TO BE USED FOR ALL TRANSMITTED AND RECEIVED MESSAGES. CHECK APPROPRIATE BOX.
ITT Transmission/Mackay, P.O. Box 11248, Raleigh, North Carolina 27604 U.S.A.

It was the happy ending of one story. It was the happy start of another.

WEATHER MAPS

THE WORLD AS SEEN THROUGH THE EYES OF A METEOROLOGIST

AND HEARD THROUGH THE EARS OF A RADIO OFFICER

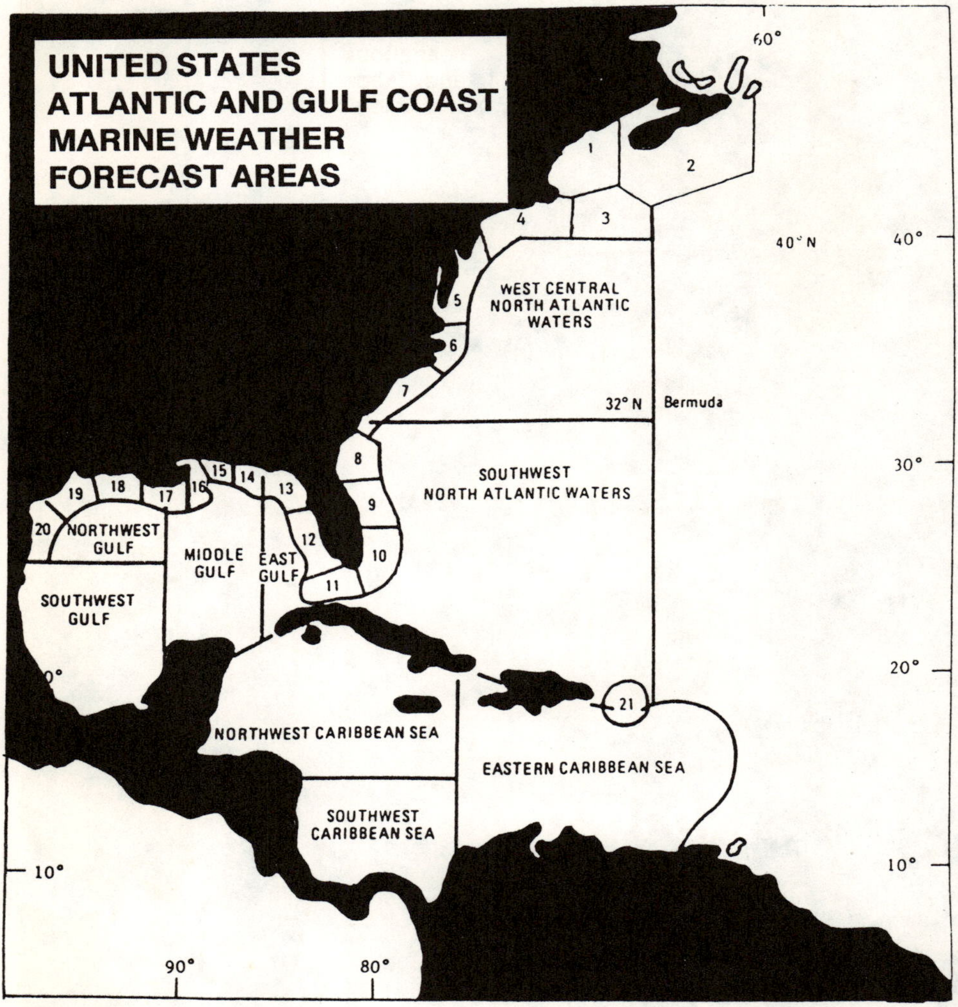
UNITED STATES
ATLANTIC AND GULF COAST
MARINE WEATHER
FORECAST AREAS
60°
1
2
3
4
40° N
40°
5
WEST CENTRAL
NORTH ATLANTIC
WATERS
6
7
32° N
Bermuda
8
SOUTHWEST
NORTH ATLANTIC WATERS
30°
9
10
11
12
13
14
15
16
17
18
19
20
NORTHWEST
GULF
MIDDLE
GULF
EAST
GULF
SOUTHWEST
GULF
20°
21
NORTHWEST CARIBBEAN SEA
EASTERN CARIBBEAN SEA
SOUTHWEST
CARIBBEAN SEA
10°
10°
90°
80°

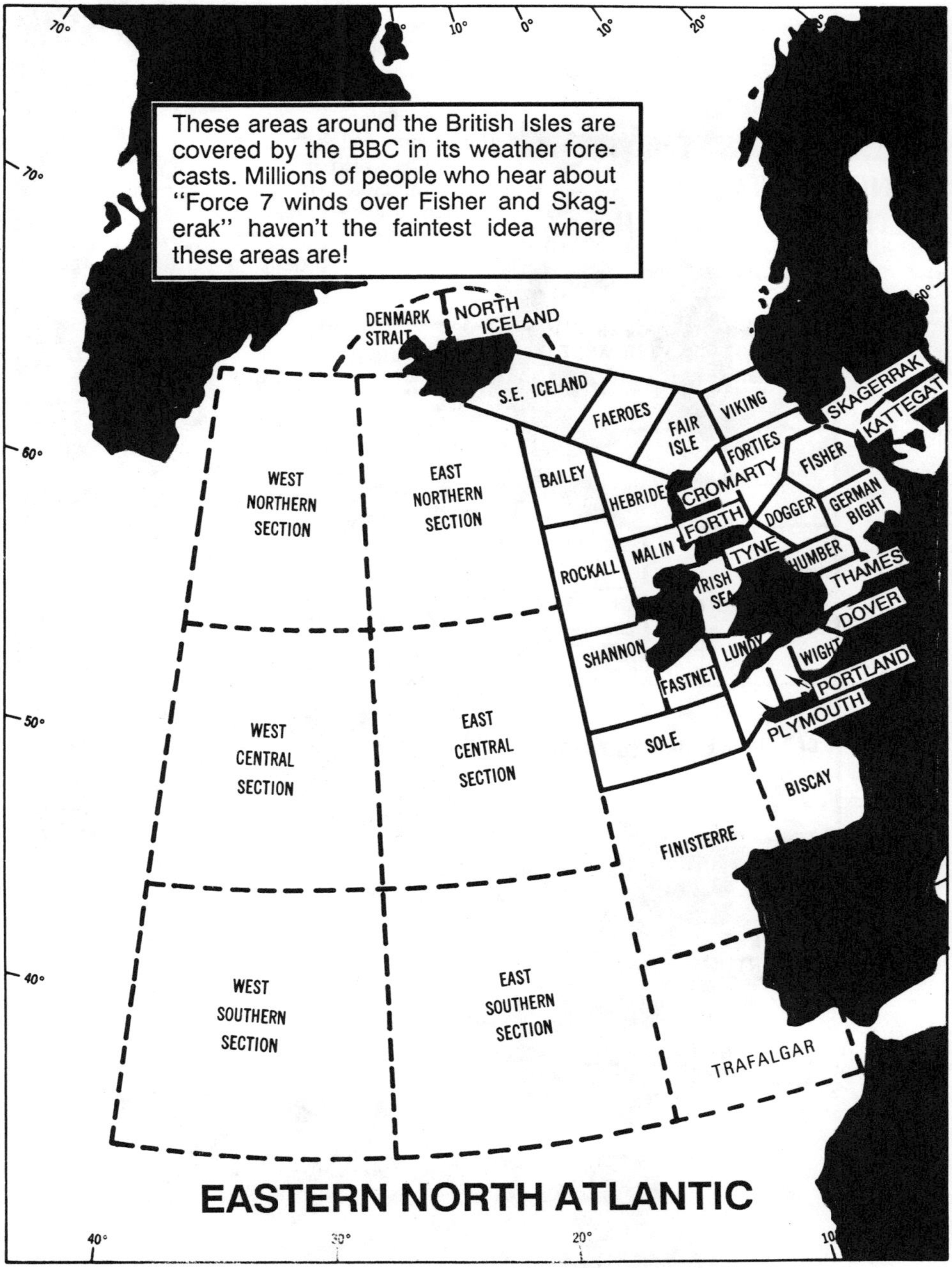
These areas around the British Isles are covered by the BBC in its weather forecasts. Millions of people who hear about "Force 7 winds over Fisher and Skagerak" haven't the faintest idea where these areas are!
DENMARK STRAIT
NORTH ICELAND
S.E. ICELAND
FAEROES
FAIR ISLE
VIKING
SKAGERRAK
KATTEGAT
FORTIES
FISHER
CROMARTY
BAILEY
HEBRIDES
FORTH
DOGGER
GERMAN BIGHT
MALIN
TYNE
HUMBER
THAMES
ROCKALL
IRISH SEA
DOVER
SHANNON
LUNDY
WIGHT
FASTNET
PORTLAND
PLYMOUTH
SOLE
BISCAY
FINISTERRE
TRAFALGAR
WEST NORTHERN SECTION
EAST NORTHERN SECTION
WEST CENTRAL SECTION
EAST CENTRAL SECTION
WEST SOUTHERN SECTION
EAST SOUTHERN SECTION
EASTERN NORTH ATLANTIC
70°
10°
0°
10°
20°
60°
50°
40°
40°
30°
20°

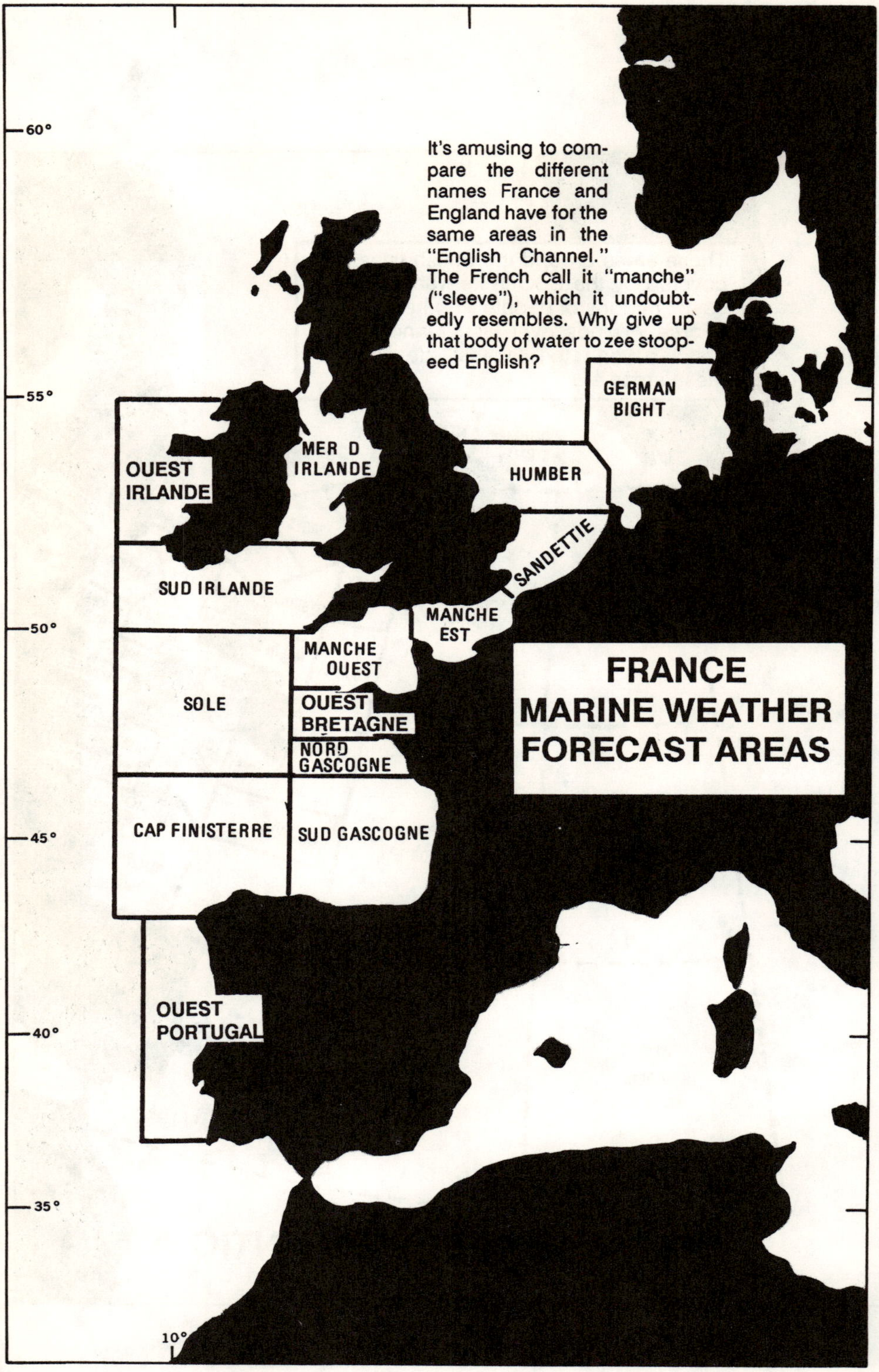
60°
55°
50°
45°
40°
35°
10°
It's amusing to compare the different names France and England have for the same areas in the "English Channel." The French call it "manche" ("sleeve"), which it undoubtedly resembles. Why give up that body of water to zee stoopeed English?
GERMAN BIGHT
HUMBER
MER D IRLANDE
OUEST IRLANDE
SANDETTIE
SUD IRLANDE
MANCHE EST
MANCHE OUEST
SOLE
OUEST BRETAGNE
NORD GASCOGNE
CAP FINISTERRE
SUD GASCOGNE
OUEST PORTUGAL
FRANCE
MARINE WEATHER
FORECAST AREAS

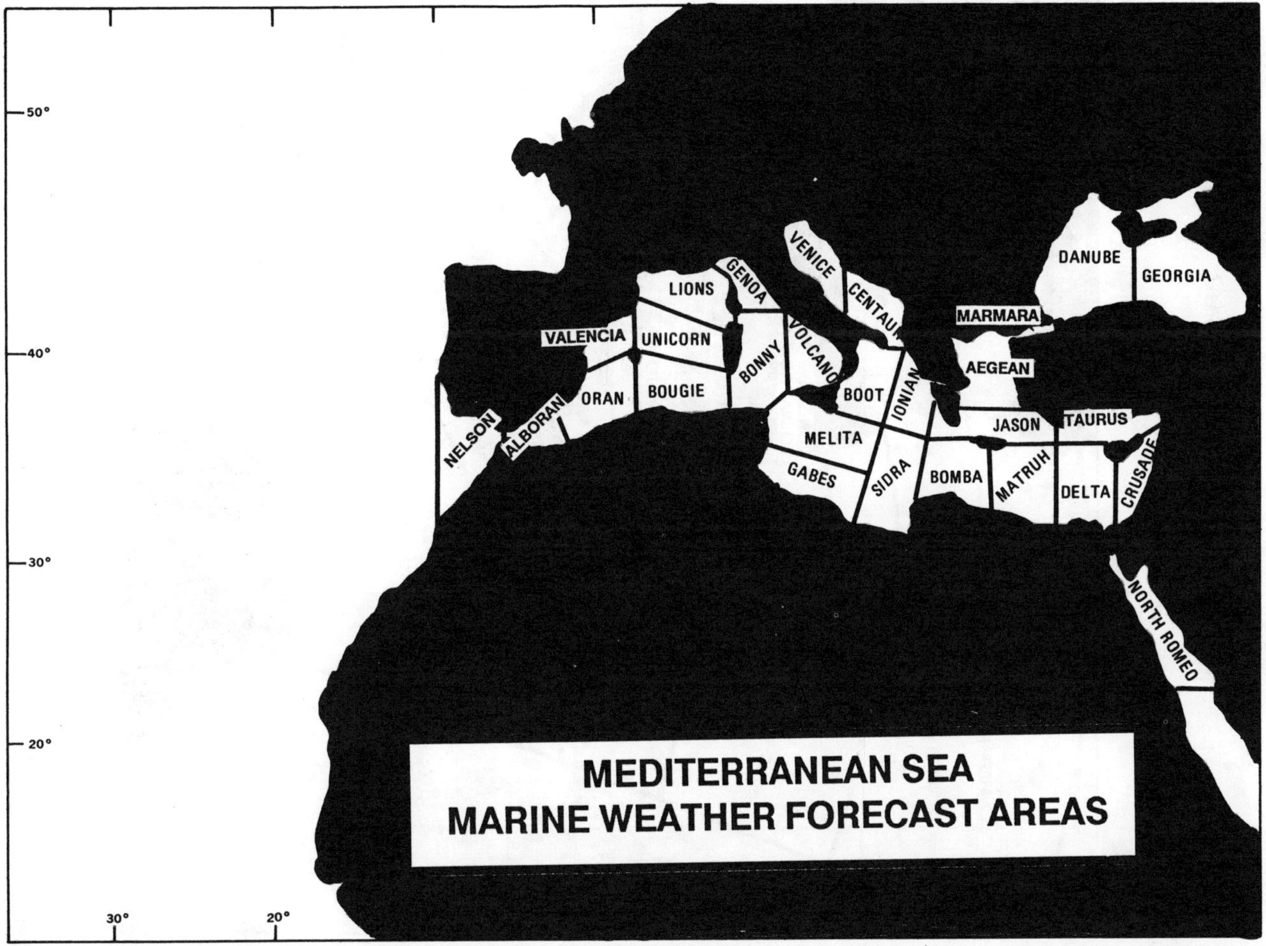
MEDITERRANEAN SEA
MARINE WEATHER FORECAST AREAS
50°
40°
30°
20°
30°
20°
NELSON
ALBORAN
VALENCIA
ORAN
LIONS
UNICORN
BOUGIE
GENOA
BONNY
VENICE
CENTAUR
VOLCANO
BOOT
IONIAN
MELITA
GABES
SIDRA
BOMBA
MATRUH
JASON
AEGEAN
MARMARA
TAURUS
DELTA
CRUSADE
DANUBE
GEORGIA
NORTH ROMEO

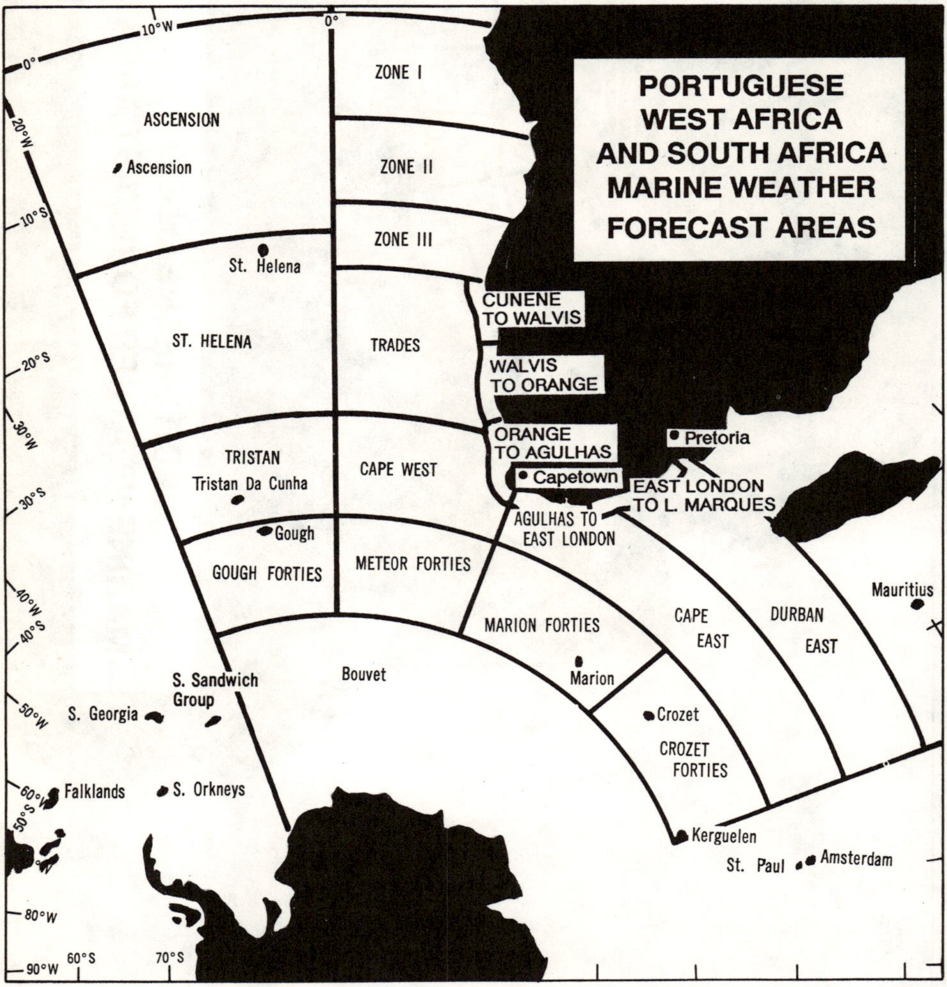

PORTUGUESE WEST AFRICA AND SOUTH AFRICA MARINE WEATHER FORECAST AREAS
10°W
0°
20°W
10°S
20°S
30°W
30°S
40°W
40°S
50°W
60°W
50°S
80°W
90°W
60°S
70°S
ZONE I
ZONE II
ZONE III
ASCENSION
Ascension
St. Helena
ST. HELENA
TRADES
CUNENE TO WALVIS
WALVIS TO ORANGE
ORANGE TO AGULHAS
Pretoria
Capetown
EAST LONDON TO L. MARQUES
AGULHAS TO EAST LONDON
TRISTAN
Tristan Da Cunha
CAPE WEST
Gough
GOUGH FORTIES
METEOR FORTIES
MARION FORTIES
CAPE EAST
DURBAN EAST
Mauritius
Bouvet
Marion
S. Sandwich Group
S. Georgia
Crozet
CROZET FORTIES
Falklands
S. Orkneys
Kerguelen
St. Paul
Amsterdam

RESCUE!

Actual names of ship & Company are withheld at Company's request.

ALL PHOTOS BY AUTHOR, WHO WAS RADIO OFFICER
(and therefore in charge of communications).

RESCUE!

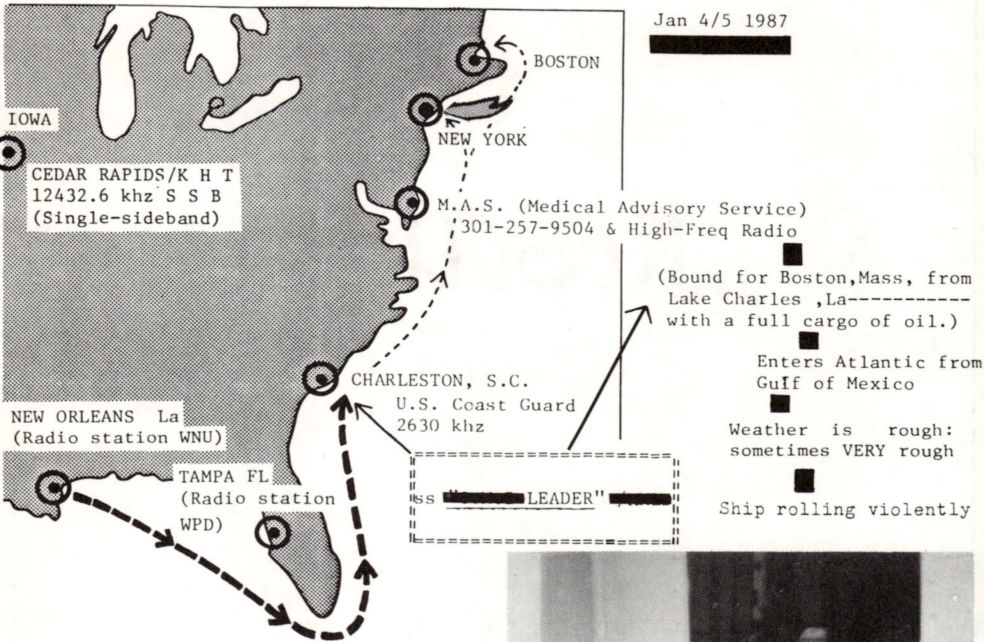

Henry L. Yeats, AB, is on watch, on wing of bridge.

He sits on tall chair.

Ship rolls, Yeats is flung to deck

He's taken below, put into his bunk, bandaged, given first aid, etc., and carefully watched, especially by Chief Mate John Millar.

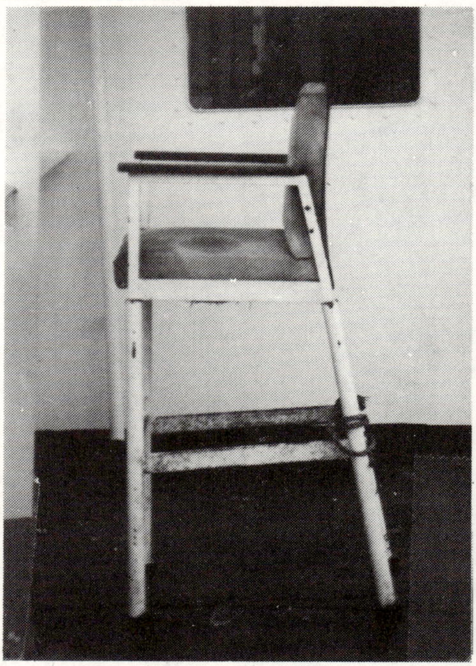

Yeats was un-watched for a few minutes. He seemed to be sleeping. When seen again, the bandages and dressings were off his head, and his wound was pumping out blood - spurting blood with a regular rhythm - an obvious arterial bleeding.

Medical advice was sought (via KHT station in Iowa, with whom ship has constant communication, mainly for contacting New York office).

They steered Capt. Ed O'Donnell to Medical Advisory Service in Maryland...Dr. Brown was given picture, details, etc., and he advised treatment and careful checking periodically of "vital signs" (pulse, breathing, temperature, etc.). Yeats was now on round-the-clock observation.

Charleston, SC U.S. Coast Guard station--and their doctors--now got into the picture. Boston, Massachusetts was a long way off. The ship might have to go into Charleston to leave Yeats under proper medical care. Contact was made with them on a medium frequency (ship was about 70 miles offshore). High-frequency contact with Cedar Rapids, Iowa and M.A.S. in Maryland was also maintained. (Voice contact, not Morse.) Morse/voice--High-Freq. single-sideband--medium-freq.--VHF for short distances--walkie-talkies--were all going to be used.

The weather was bad and getting worse. Capt. O'Donnell proposed heading for Charleston, SC and getting inside the breakwater--calm water--before trying to transfer Yeats to launch. Charleston agents asked to cooperate.

Radiogram Via ITT RS79

ITT Mackay — A Division of ITT Telecommunications Corporation

All messages are accepted subject to rates, rules and regulations in the applicable tariffs on file with the Federal Communications Commission/I.T.U

Prefix: MSG No: 1
Check: 30
Date: JAN 5

Office of Origin: [redacted] LEADER
Time Filed: 1459 GMT

Charges
Ship
Coast
Landline
Tax U.S.
Misc. ~~10.13~~
Total

QRC US02

(agent)

CAROLINE CHARLESTONSC (TLX 810-881-1734)

PLEASE COORDINATE WITH MEDICAL ADVISORY SYSTEM PORTEVERGLADES AND USCG TO EVACUATE INJURED ABLE-SEAMAN . REQUEST PILOT TO EFFECT TRANSFER IN CALM WATER

REGARDS

MASTER

(New Orleans.Slidell-radio)

*Received ☐ Sent ☒ WNU (Station) Date: 1/5 Time: 1525GMT (GMT) Signed:

*THIS FORM IS TO BE USED FOR ALL TRANSMITTED AND RECEIVED MESSAGES. CHECK APPROPRIATE BOX.
ITT Transmission/Mackay, P.O. Box 11248, Raleigh, North Carolina 27604 U.S.A.

The bad weather got worse--then worst. So bad Capt. O'Donnell decided to "heave to," and ride out the storm. The ship was rolling, pitching up, down, every which way, but going nowhere.

Everyone concerned with the ship had to be told.

[redacted] NEWYORK — Head Office

L A N G H L VIRGINIA — Charterer

HAVE DIVERTED SHIP TO CHARLESTONSC WITH INJURED SEAMAN ON ADVICE OF DR BROWN OF MEDICAL ADVISORY SYSTEM OF PORTEVERGLADES . AT PRESENT WE ARE HOVE TO IN VERY ROUGH SEAS 70 MILES SOUTHEAST OF CHARLESTON REGARDS

MASTER

Agent — MORANCO BOSTON

AM HOVE TO IN ROUGH WEATHER S.E. OF CHARLESTON WILL GIVE YOU NEW ETA AS SOON AS POSSIBLE

REGARDS

MASTER

via W P D/Tampa FL. CW(Morse) messages

FORE and . . .

The night wore on...vessel went up and down.
Yeats was watched--checked and checked.

(This is where the helicopter <u>might</u> try to drop/pick up someone)

<u>Time</u>: 3 AM
<u>Patient</u>: Condition not good (coughing blood). <u>M.A.S.</u> advice sought.

Dr. Brown has been replaced by female colleague, Dr. Fisher. She gets symptoms, etc., and is alarmed. She wants patient
OFF!
<u>SOONEST!</u>
by <u>HELICOPTER!</u>

Capt. O'Donnell, clinging to desk to stay stable as he talks into phone (anything not anchored will slide or bounce or noisily hit the deck), has two-word reaction to this "helicopter" idea...

NO WAY!

The U.S. Coast Guard, Charleston, come back. They want ship's exact position.

"We're sending helicopter. We'll be over you in 30 minutes."

Capt. O'Donnell sees high seas flooding the foredeck <u>and</u> after-deck...the violent rolling and pitching continues...there's just no room for anyone to land...the ship is full of flammable oil...he's responsible for the lives and safety of 25 crew members (including himself) plus two women passengers (his own wife and that of Second Mate). He looks again at MOUNTAINOUS seas and reduces two words to one:

IMPOSSIBLE!

That pig-headed U.S. Coast Guard won't be put off. We-e-ll (thinks O'Donnell)--if we change course--turn into the wind--maybe the seas on the after-deck will subside a little. And just about this time, the weather decides to improve a <u>little</u>--not much, but the waves aren't so high.

USCG "whirlybird" out of order...no sweat...machine and personnel borrowed from U.S. Marines...in 30 minutes (as promised) helicopter arrives...(bigger than expected: size of small bus).

U.S. Marines paramedic is swung aboard on a long cable

(regretfully, no pictures)

While he examines Yeats, helicopter hovers overhead

RESCUE!

U.S. Marines paramedic (in tights like a ballet dancer, a model of EFFICIENCY, appearance and manner) inspects and instructs. He and Yeats are going to cling together, be whirled upwards on a rope dangling from helicopter.

Assembly on poop-deck (right over propeller)

Yeats must be full of misgivings, as the medic explains, adjusts, tightens, etc. But he's the patient, so he's gotta be patient.

Bos'n

Yeats' fellow AB's watch, ready to help if needed

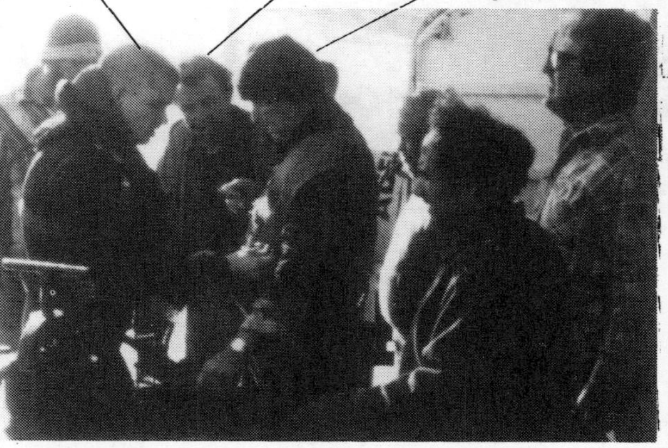

What is the Chief Mate looking at so intently?

The eager-to-help sailors watch...dash...scamper...reach up--but they can't catch the damned rope. Overhead, the helicopter pilot is doing a masterly job, holding the plane stationary. The ship rises and falls...sailors scamper...the damned rope is going every which way. The wind blows hard and keeps the rope inaccessible. But the helicopter waits patiently until everybody and everything else gets sorted out.

THIS.

See that little dot in the sky, below the helicopter? That's the end of the rope from the helicopter.

An old recipe for making rabbit pie, started, "First, catch your rabbit." Here, we must get hold of that rope end...

It goes this way...that ...forward...back... sweeps in from the left...then the right.

How come they can now stand on poop-deck? (See Page 365)

(It was IMPOSSIBLE.) Because Capt. O'Donnell had bright idea of steering ship INTO wind and sea.

Here it comes, swooping down on the starboard...

will it...

can we...

NO.

It's on the port side now.

Steady...

steady...

GOT IT!! Fix...adjust...steady...

ALLEZ--

OOOP!

RESCUE!

Swinging OUT...

and UP...

They're ready for him up there...

Robert Newby, A.B. ("horse") watches as his shipmate goes the last few feet UP...

He's pulled into the aircraft, bound for first-class medical care.

Radiogram Via ITT RS79

ITT Mackay — A Division of ITT Telecommunications Corporation
All messages are accepted subject to rates, rules and regulations in the applicable tariffs on file with the Federal Communications Commission/I T U

Prefix: MSG No: 1 | Office of Origin: NEWYORK | Charges
Check: 87/82 | | Ship
Date: Jan 7 | Time Filed: 1615 gmt | Coast
Landline
Tax U.S.
Misc.
Total

MASTER ███ LEADER/KMLD ███

A01/OMI 8264
RE AB HENRY L.YEATS, ███ LEADER VOY 200,D1/1/4/87,

OUR FILE 0078 . CHRISTA BALLANTINE OF LAMORE BURNES

REPRESENTING OUR UNDERWRITERS WILL BOARD DURING YOUR

FORTHCOMING CALL AT BOSTON TO INVESTIGATE ABOVE INJURY

PLEASE FULLY COOPERATE WITH HER. MR YEATS SEEMINGLY SUFFERED

SUBDURAL HAEMATOMA,BUT IS IN STABLE CONDITION IN ST.FRANCIS

XAVIER HOSPITAL IN CHARLESTON.WIFE INFORMED.DOCTOR'S OPINION,

HELICOPTER EVACUATION PROBABLY SAVED LIFE. THANKS & REGARDS

MCDONALD,CLAIMS ███

* Received X Sent ☐ WPD (Station) Date: 1/7 Time: 1700 z (GMT) Signed:

*THIS FORM IS TO BE USED FOR ALL TRANSMITTED AND RECEIVED MESSAGES. CHECK APPROPRIATE BOX.
ITT Transmission/Mackay, P.O. Box 11248, Raleigh, North Carolina

H A P P Y E N D I N G

Radiogram Via ITT RS79

ITT Mackay — A Division of ITT Telecommunications Corporation
All messages are accepted subject to rates, rules and regulations in the applicable tariffs on file with the Federal Communications Commission/I T U

Prefix: MSG No: 1 | Office of Origin: NEWYORK | Charges
Check: 69/65 | | Ship
Date: 14 | Time Filed: 2143gmt | Coast
Landline
Tax U.S.
Misc.
Total

MASTER ███ LEADER ███

A 01/OM I8850. RE HENRY YEATS. HELICOPTER EVACUATION VOY 200.

OUR FILE 0078. RECEIVED WITH THANKS REPORTS AND PHOTOGRAPHS RE. ABOVE

INCIDENT. HELICOPTER COMMANDER DAVID KENNEDY PARTICULARLY INTERESTED

IN OBTAINING COPIES OF PHOTOGRAPHS. PLEASE SEND MY ATTENTION DUPLICATE

SET OF PRINTS OR NEGATIVES IF MORE CONVENIENT. FYI YEATS FEELING GOOD

BUT STILL WEAK AND REMAINING HOSPITALIZED FEW MORE DAYS.

THANKS RGDS

..CDONALD CLAIMS ███

* Received ☐ X XX Sent ☐ WPD (Station) Date: 1/15 Time: 1354z (GMT) Signed:

*THIS FORM IS TO BE USED FOR ALL TRANSMITTED AND RECEIVED MESSAGES. CHECK APPROPRIATE BOX
ITT Transmission/Mackay, P.O. Box 11248, Raleigh, North Carolina 27604 U.S.A.

CORP. PARK AVENUE, NEW YORK, N.Y. 10017

Leader
Corp.
Park Ave,
New York City, N.Y.

Jan.11,1987

Dear Sirs,

On Jan.5th, off the coast of Charleston, S.C. the USCG in coordination with the US Marines effected a rescue operation by removing an injured seaman from my vessel by helicopter. I have since learned that without this effort on the part of the USCG and US Marines the seaman might possibly have died. I wish to thank and commend the personnel involved. I did not think it possible for a helicopter to do rescue work in the high winds involved. Add to that the rolling and pitching of the ship and I was convinced it wouldn't work. However the Pilot of " Angel One" showed remarkable skill in landing a Medic and then picking him and the patient up. I am thankful all went so well. As a seaman I am gratified to know such fine men are available to help us when needed.

Thanks again,

E.M.O'Donnell, Master, LEADER

CABLE NEW YORK TELEX

Giving Credit Where Due . . .

- The spectacular helicopter feat by the US Coast Guard and US Marines — with the weather, Captain O'Donnell (at first) and sheer common sense all opposing — calls for our admiration and applause . . .
- What happened AFTER all this (the opinion of the Charleston, SC hospital that it had been "just in time" to save Yeats' life) confirms and bolsters this applause.
- What happened BEFORE — Chief Mate Millar's careful watch over Yeats' "vital signs" (breathing, temperature, blood pressure, etc.) — in obedience to the medical advice he'd been given — had this importance: without such attention, Yeats might well have died.
- Captain O'Donnell . . .
    - instigating the calls for medical assistance,
    - maneuvering his ship to make the rescue possible,
    - and his overall control of their situation, to make the various factors and people concerned cooperate fully.

Without all this, the factors/people might have clashed or gone their own independent ways . . .

and Yeats would have died.

- The doctors . . .
    - advising at first from a distance,
    - then treating their patient directly, were, of course, obvious cause of Yeats' survival.

- ALL THESE PEOPLE DEPENDED ON THE SAME THING — CONTACT BY RADIO. INTERCHANGE OF INFO . . . BY VOICE OR MORSE
    - FROM A THOUSAND MILES VIA SINGLE-SIDEBAND RADIO
    - TO 70 MILES BY MEDIUM-WAVE TRANSMITTERS RECEIVERS
    - TO 100 FEET BY VHF AND WALKIE-TALKIE RADIOS (FOR HELICOPTER/SHIP CONVERSATIONS).
- THIS IS THE VITAL FACTOR ON WHICH EVERYTHING/EVERYONE DEPENDED.

WITHOUT IT, THERE IS NO "MAYBE."
YEATS WOULD HAVE DIED.

A chain is as strong as its weakest link. All links, here, were STRONG.

BIBLIOGRAPHY

My Father Marconi, by Degna Marconi (New York: McGraw-Hill, 1962)

The Night Lives On, by Walter Lord (New York: William Morrow & Co., 1986)

Oceanus (Woods Hole: Woods Hole Oceanographic Institute, 1985)

Origin of Maritime Radio, by R.F. Pocock and G.R.M. Garrett (H. M. Stationery Office, 1972)

SOS: The Story of Radio Communication, by G.E.C. Wedlake (Newton Abbott: David & Charles, 1973)

SOS To The Rescue, by Karl Baarslag (New York: Oxford University Press, 1935)

The Strange Fate of the Morro Castle, by Gordon Thomas and Max Morgan Witts (Stein and Day, 1972)

A WORD OF SPECIAL THANKS

As this book was going to press, I realized that I had made a real blooper. It was a sin of omission, failing to thank the Society of Wireless Pioneers for their help in the research and promotion of this book. These few paragraphs are scant praise for all their help to me.

The SOWP was founded in 1968, the brainchild of Bill Breniman, who has also been its main driving force and super-organizer. For 20 years the SOWP has put itself at the very center of radio-communication history, preserving records, memorabilia and historical facts concerned with radio telegraphy. Their research facilities in this field are unequalled.

The SOWP has been mentioned a few times in this book. There is great camaraderie and exchange of information.

SOWP has enrolled more than 5,000 members since 1968 and is still growing. *The Sparks Journal* and other publications keep members enthusiastic about their important role in 20th-century communications. *QTC* would have been a much poorer book without all the groundwork and research supported and encouraged by the SOWP. The 20th century (especially the first half) belonged to Sparks. This legacy will be preserved into the 21st and 22nd centuries by the mighty labors of the SOWP. Hats off and hurrah to them.

Interested radio professionals may contact Paul Dane, Executive Director, 146 Coleen Street, Livermore, CA 94550.